Johannes Held

Non-Supersymmetric Flux Compactifications

Johannes Held

Non-Supersymmetric Flux Compactifications
of Heterotic String- and M-theory

Südwestdeutscher Verlag für Hochschulschriften

Impressum/Imprint (nur für Deutschland/only for Germany)
Bibliografische Information der Deutschen Nationalbibliothek: Die Deutsche Nationalbibliothek verzeichnet diese Publikation in der Deutschen Nationalbibliografie; detaillierte bibliografische Daten sind im Internet über http://dnb.d-nb.de abrufbar.
Alle in diesem Buch genannten Marken und Produktnamen unterliegen warenzeichen-, marken- oder patentrechtlichem Schutz bzw. sind Warenzeichen oder eingetragene Warenzeichen der jeweiligen Inhaber. Die Wiedergabe von Marken, Produktnamen, Gebrauchsnamen, Handelsnamen, Warenbezeichnungen u.s.w. in diesem Werk berechtigt auch ohne besondere Kennzeichnung nicht zu der Annahme, dass solche Namen im Sinne der Warenzeichen- und Markenschutzgesetzgebung als frei zu betrachten wären und daher von jedermann benutzt werden dürften.

Coverbild: www.ingimage.com

Verlag: Südwestdeutscher Verlag für Hochschulschriften GmbH & Co. KG
Heinrich-Böcking-Str. 6-8, 66121 Saarbrücken, Deutschland
Telefon +49 681 37 20 271-1, Telefax +49 681 37 20 271-0
Email: info@svh-verlag.de

Approved by: München, LMU, Diss, 2012

Herstellung in Deutschland (siehe letzte Seite)
ISBN: 978-3-8381-3342-3

Imprint (only for USA, GB)
Bibliographic information published by the Deutsche Nationalbibliothek: The Deutsche Nationalbibliothek lists this publication in the Deutsche Nationalbibliografie; detailed bibliographic data are available in the Internet at http://dnb.d-nb.de.
Any brand names and product names mentioned in this book are subject to trademark, brand or patent protection and are trademarks or registered trademarks of their respective holders. The use of brand names, product names, common names, trade names, product descriptions etc. even without a particular marking in this works is in no way to be construed to mean that such names may be regarded as unrestricted in respect of trademark and brand protection legislation and could thus be used by anyone.

Cover image: www.ingimage.com

Publisher: Südwestdeutscher Verlag für Hochschulschriften GmbH & Co. KG
Heinrich-Böcking-Str. 6-8, 66121 Saarbrücken, Germany
Phone +49 681 37 20 271-1, Fax +49 681 37 20 271-0
Email: info@svh-verlag.de

Printed in the U.S.A.
Printed in the U.K. by (see last page)
ISBN: 978-3-8381-3342-3

Copyright © 2012 by the author and Südwestdeutscher Verlag für Hochschulschriften GmbH & Co. KG and licensors
All rights reserved. Saarbrücken 2012

Contents

Abstract   vii

1 Introduction   1
    1.1 The supergravity approach to string theory 2
       1.1.1 String theory and its low energy limit 2
       1.1.2 Advantages of the supergravity approach 7
    1.2 The moduli problem in supergravity 9
       1.2.1 Calabi-Yau compactification and moduli 10
       1.2.2 Stabilizing moduli with flux 11
    1.3 G-structures and flux compactifications 12
    1.4 Outline and results 15

2 Heterotic Supergravity Theories   17
    2.1 Eleven-dimensional supergravity 17
       2.1.1 The bulk action 18
       2.1.2 Effects of the boundary 19
    2.2 Heterotic supergravity 21
    2.3 Dimensional reduction of M-theory 22
    2.4 Scalar potential and equations of motion 27
       2.4.1 Heterotic M-theory 27
       2.4.2 Scalar potential of heterotic supergravity 30

3 G-Structures   37
    3.1 General remarks on G-structures 37
       3.1.1 An intuitive definition of G-structures 37
       3.1.2 G-structures and torsion 40
       3.1.3 G-structures and supersymmetry 41
    3.2 SU(3) structures in six dimensions 42

		3.3	G_2 and SU(3) structure in seven dimensions	44

4 Heterotic Domain Wall Supersymmetry Breaking   49
    4.1   BPS-like potential and SUSY conditions 50
    4.2   Supersymmetry breaking vacua: general discussion 55
           4.2.1   Torsion induced SUSY-breaking vacua 55
           4.2.2   Gravitino and gaugino mass 61
           4.2.3   Conditions on the curvature 63
    4.3   NS5-branes, calibrations and bundle stability 65
    4.4 $\frac{1}{2}$ Domain-Wall supersymmetry breaking 70
           4.4.1 $\frac{1}{2}$ DWSB vacua 70
           4.4.2   Four-dimensional interpretation 73
    4.5   Examples via homogeneous fibrations 76
           4.5.1   Constraints on the elliptic fibration 77
           4.5.2   Simple examples 85
    4.6   Adding a gaugino condensate 88
           4.6.1   Gaugino condensate and no-scale SUSY-breaking 88
           4.6.2   Supersymmetric AdS_4 vacua and calibrations 92
           4.6.3 $\frac{1}{2}$ DWSB AdS_4 vacua with gaugino condensate 94

5 BPS-Potentials in M-Theory   99
    5.1   The Ricci scalar of G_2 manifolds 100
           5.1.1 R in terms of G_2 structure 101
           5.1.2 R in terms of SU(3) structure 101
    5.2   Supersymmetry conditions 102
           5.2.1   Differential conditions 103
           5.2.2   Conditions on the flux 104
    5.3   Is a BPS-like potential possible? 105
           5.3.1   Bulk potential 105
           5.3.2   Boundary potential 110
    5.4   Limiting cases 112
           5.4.1 G_2 holonomy 112
           5.4.2   SU(3) holonomy 114
           5.4.3   The ten-dimensional limit 115
    5.5   DWSB in heterotic M-theory? 117

6 Conclusion   121

A Appendix 125
- A.1 Conventions 125
- A.2 Dual formulation of heterotic supergravity 129
- A.3 Supersymmetry breaking in the presence of a gaugino condensate 130
- A.4 The scalar curvature of G_2 structure manifolds 132
- A.5 SUSY constraints 135
- A.6 SUSY conditions for eleven-dimensional SUGRA without boundary 136

List of Figures

4.1 BPS objects of the theory, in terms of their four-dimensional appearance. 66

List of Tables

A.1 Constraints from $\delta\Psi_\mu = 0$ coming from $\eta_+^T \gamma_{[n]} \text{Ext} + \text{Ext}^T \gamma_{[n]} \eta_+$. 135
A.2 Constraints from $\delta\Psi_\mu = 0$ coming from $\eta_+^T \gamma_{[n]} \text{Ext} - \text{Ext}^T \gamma_{[n]} \eta_+$. 135
A.3 Constraints from $\delta\Psi_\mu = 0$ coming from $\eta_+^\dagger \gamma_{[n]} \text{Ext} + \text{Ext}^\dagger \gamma_{[n]} \eta_+$. 135
A.4 Constraints from $\delta\Psi_\mu = 0$ coming from $\eta_+^\dagger \gamma_{[n]} \text{Ext} - \text{Ext}^\dagger \gamma_{[n]} \eta_+$. 136

Abstract

This dissertation is concerned with non-supersymmetric vacua of string theory in the supergravity (SUGRA) approach. This approach is the effective description of string theory at low energies.

The concrete field of research that is treated here is heterotic $E_8 \times E_8$ string theory at weak and at strong coupling, respectively. In the strong coupling limit the theory is described by eleven-dimensional SUGRA with two ten-dimensional boundaries (heterotic M-Theory). The transition to the weak coupling limit is governed by the restricted space dimension, whose length tends to zero for weak coupling such that the two boundaries get identified with each other. The resulting theory is ten-dimensional $E_8 \times E_8$ SUGRA.

In the context of this heterotic SUGRA, at first six of the former nine space-dimensions are compactified, and then, in the presence of non-vanishing background flux, conditions for unbroken supersymmetry (SUSY) in four space-time dimensions are analyzed. Afterwards, a violation of one of the necessary SUSY conditions is allowed. An essential ingredient, necessary for this to work, is the presence of flux. This kind of SUSY-breaking leads to severe constraints on the compact six-dimensional manifold, which can be satisfied by fiber bundles with two-dimensional fiber and four-dimensional base. In simple examples one can stabilize the expectation value of the dilaton as well as the volume of the fiber, whereas the volume of the base remains undetermined.

Furthermore, the effect of a fermionic condensate is analyzed. The expected additional SUSY-breaking can be observed, and it is shown that the breaking induced by the flux can not be canceled by the contributions from the condensate.

The end of this thesis is concerned with the discussion of the strong coupling limit of the previously found examples. To analyze this, it is necessary to rewrite the action of heterotic M-theory as a sum of quadratic terms, which vanish once SUSY is imposed. However, as is clarified by a detailed analysis, this is only possible for certain cases.

Nonetheless, one can show the consistency of the M-theoretic results with the findings in the weak coupling limit. Moreover, it is proofed that it is possible to treat the class of examples constructed in heterotic SUGRA also in the strong coupling limit.

Chapter 1

Introduction

The results of our works [1–4] could be summarized in the following concise way: 'In this thesis we are concerned with the problem of constructing non-supersymmetric flux compactification vacua in the context of heterotic string theory and the lift of eventually found vacua to eleven-dimensional supergravity with boundaries'. Such a statement is of course not very self-enlightening and is in need for a further clarification and a localization in the context. This is the goal of this introductory chapter.

This thesis uses as its basic framework the vast topic of string theory. In this context we have derived some new results on supersymmetry breaking in heterotic string compactifications. Moreover, the thesis represents to a good extent working techniques that are essential to understand a wide range of publications. Therefore, by explaining how this thesis is connected to string theory we will also shed light on more general trends in this field of research. On the other hand, such an explanation will also prepare the reader for the more technical details that are comprised in the main part of this work. Last but not least it will in addition relate the results of this thesis to recent developments in string theory.

Thus, a natural question to start with is to ask which working techniques we employ. We therefore have to make clear that our whole discussion will take place in the language of supergravity theories. Moreover, we will work in a classical regime, i.e. we will consider only background solutions. As we think, this is in need of an explanation, and we will therefore try to make clear the connection between string theory and supergravity in section 1.1. During this discussion we will also show why eleven-dimensional supergravity has to be viewed as a part of string theory.

The second important point to mention is that we are dealing with so-called flux-compactifications. We will therefore discuss in section 1.2 the main problem that first led to the inclusion of non-vanishing fluxes. In fact, all supergravity approaches to string theory suffer from massless scalar fields, called moduli, which have to be absent in phenomenological viable models. One mechanism to give a mass to some (or even all) of these moduli is to use backgrounds with non-zero flux.

This leads us directly to the recent developments in the field of flux-compactification, that we will shortly discuss in section 1.3. Here, we will focus on the use of the mathematical framework of G-structures and on the construction of non-supersymmetric vacua, since these are the starting points for our own work. At the end of this chapter we will comment on the organization of this thesis and summarize our main results.

1.1 The supergravity approach to string theory

In this section we are going to describe the supergravity (SUGRA) approach to string theory. As it can be understood as the low energy limit of string theory, it will be inevitable to review some basic features of string theory as well. However, on this we will be as brief as possible and refer the reader to the well known textbooks on the topic [5–10] for further information.[1] Besides introducing SUGRA, we will also discuss the advantages of the approach.

1.1.1 String theory and its low energy limit

World sheet theory

String theory is in its essence the attempt to do high energy physics with one-dimensional objects instead of point particles. It was developed at the end of the 1960's in order to deal with the strong interaction, but was discarded after the development of quantum chromodynamics (QCD) (for a history of early string theory see e.g. [5, 20–22] and references therein). However, since there appeared states with spin $s = 2$ in the spectrum, and since the mediating particle of gravitation, the graviton, has also spin $s = 2$, string theory was revitalized in the mid eighties as a possible theory of quantum gravity [23–27].

The action of a freely moving point-particle can conveniently be described by the integral over the world-line of the particle (i.e. the path of the particle in space-time). The basic idea of string theory is to replace this action by the integral over the two-dimensional surface swept out by a one-dimensional object in space-time, the string. This gives a parametrization of the space-time coordinates in terms of two world-sheet coordinates. A different view on string theory is therefore to consider it as a two-dimensional theory with a number of bosonic fields, which equals that of the space-time dimension.

In order to make the action dimensionless one has to introduce one length scale l_s, that can be understood as the typical length of a string. This is the only dimensionful parameter that has

---

[1] One should note that due to the limited space, but also due to the minimal connection to the works presented here, we will omit several research directions in our discussion. The most important ones of these are the AdS/CFT correspondence [11], F-theory [12], the string landscape [13], string inflation [14], black hole physics [15], doubled field theory [16,17], and non-geometric fluxes [18,19].

to be introduced by hand into the theory and it implies that massive states of string theory have masses of order $1/l_s$. In general, since one wants to describe quantum gravity with string theory it is reasonable to identify the string scale $1/l_s$ roughly with the Planck scale $M_p \cong 10^{19}$ GeV. However, there are also models which make it possible to have a lower string scale [28, 29].[2]

As it turns out, this two-dimensional description of string theory has a conformal symmetry[3] promoting it to a conformal field theory (CFT). Starting from there, it is possible to deduce a number of strong results. First, in order for the spectrum not to contain a tachyon it is mandatory to add a second sector to the theory such that the world sheet theory becomes supersymmetric. However, due to consistency at the one loop level [23], it turns out that states with the same mass appear as supersymmetry (SUSY) multiplets also from the space-time perspective. Pointed out more concisely: string theory without tachyons has to obey supersymmetry.

Quantization of a field theory brings with it the danger that not all symmetries which are preserved at tree level can automatically be kept at one-loop level. These so-called anomalies will effectively destroy the symmetry properties, and thus render the theory inconsistent (see e.g. [39, 40] and references therein). Anomalies also occur during the quantization of the world-sheet theory of string theory. However, it can be shown that they are absent when one fixes the number of bosonic fields to be ten.[4] Since the number of bosonic fields equals the number of space-time dimensions, superstring theory can only be formulated in ten dimensions. The same analysis which gives the dimensional constraint yields also equations which involve the space-time curvature and can be interpreted as equations of motion (EoM's) of a low-energy space-time theory [41–44].[5] This is a first hint of how to obtain a low energy limit of string theory.

Satisfying these conditions it turns out that there are only five consistent ways to formulate string theory. The so-called type II theories contain only closed strings. There are two of them that differ in its spectrum, type IIA which is non-chiral, and type IIB which is chiral [45, 46]. Type I string theory also contains closed strings, but has the additional requirement that all word-sheets that are swept out by strings have to be non-orientable. For this reason it is also possible to include open strings to the spectrum [47, 48]. The last class of theories, which will concern us mostly in this thesis, are the heterotic string theories. In these, the closed strings can also carry gauge charges, while open strings are forbidden. One can then distinguish two heterotic theories by the allowed gauge groups, which are either SO(32) or $E_8 \times E_8$ [25, 26].

---

[2]The implication of a low string scale on LHC physics has been analyzed in [30–32]. Experimentally excluded is the energy region below 1.67 TeV [33].

[3]Roughly speaking, a conformal symmetry leaves the angles between two lines equal. For more information see [34–38].

[4]By 'bosonic fields' we mean here the fields which describe the embedding of the string world sheet into space-time, and not other bosonic states as the dilaton or the Neveu-Schwarz flux.

[5]These equations appear as beta-functions of the world-sheet theory that have to vanish in order for the theory to be anomaly free.

Conceptually important in the above description of string theory is that it is defined intrinsically as a power series. The expansion parameter in this series, the string coupling g_s can be traced back to the topology of the string world sheet, counting the number of handles it has. In this sense, it is analogous to a loop expansion in quantum field theory and is only sensible as long as the coupling g_s is small. However, this can not be guaranteed from the beginning, since g_s is determined by the theory itself [5–10]. Therefore, at this point one could wonder how the theory is defined in the strong coupling regime, or whether it can be defined at all.

Branes and dualities

The above description reflects the state of the art before the mid 1990's. However, then two new concepts were introduced that revolutionized string theory. On the one hand, it was discovered that there can be other extended objects besides strings, which can be included within the theory. These fill $(p+1)$-dimensional space-time, and since strings can end on them and obey Dirichlet boundary condition at the endpoints, they were named Dp-branes [49–51]. On the other hand, it was recognized that all of the five formulations of the theory are actually related to each other by so-called dualities [52–58].

The simplest of these dualities goes under the name of T-duality [59–62]. To understand its origin it is necessary to imagine that one could role up at least one dimension of space as a circle of radius R. This is actually an old idea, which was promoted by Kaluza and Klein in the 1920's [63, 64] in order to unify gravity and electromagnetism, but was discarded due to its predictions: a tower of massive states with even mass separation of the order of the inverse radius $1/R$, and a scalar field. However, since string theory is only consistent in ten dimensions, it is more or less unavoidable to have compact space dimensions if the theory should be able to describe physics in four dimensions.

The implications of these extra dimensions for string theory are quite interesting. In contrast to a point particle, a closed string can not only propagate in the compact dimension, but can also wind around it such that it is not contractible any more. The mass of such a string, when viewed from the perspective of the non-compact dimensions, has two extra contributions: one from the quantized momentum in the internal dimension, which would also be present for a point particle, and one being proportional to the number of times the string is wrapped around the extra dimension. The important point to notice is that the mass contribution from the momentum is proportional to an integer times $1/R$, while the winding contribution is proportional to the radius R (the larger R the larger is the tension in the string and hence its mass). Going to the limit of infinite radius, i.e. to the uncompactified theory, one gets back a continuous (and therefore propagating) momentum, and decouples the infinitively heavy winding modes. On the other hand, taking the radius to zero decouples the momentum modes, but renders the winding modes continuous. Physically one cannot distinguish these two regimes. In both cases one has a continuous mass spectrum for every spatial dimension and the fact that in the limit $R \to 0$ it is produced by winding modes cannot be detected anymore.

It is therefore sensible to argue that the theory at radius R is dual to the theory at radius $1/R$, where 'dual' means that both describe the same physics. This is essentially what T-duality does. Extending the duality to the open string leads to even more interesting results. In fact, an open string cannot wind anything as long as its endpoints are not fixed. So what T-duality claims is that a string that moves freely in all space dimensions on the one side of the duality is stuck to a fixed plane on the other side. These planes are the aforementioned D-branes. This means that since T-duality is valid in the closed string sector, one has to introduce D-branes if one introduces open strings [49–51].

Analyzing these issues more carefully one can show that several of the above five descriptions of string theory are connected to each other by T-duality. Starting from the type I string, one reaches the two type II theories by shrinking an odd (type IIA) or an even (type IIB) number of space dimension to zero and going to the T-dual description. This in turn implies that the type II theories are mutually T-dual to each other. Furthermore, since the open strings of type I can end everywhere in space, there has to be a D9-brane present. Thus, type IIA theory has only Dp-branes with even p, whereas type IIB contains only branes with odd p. Hence, when T-duality is included type I and the type II string theories should be viewed as one single theory [53].

In a similar fashion one can argue that the two heterotic theories are connected by a T-duality and provide in fact also only one description of string theory [52]. However, in the heterotic case there are no D-branes, as there are no open strings. But another type of branes, called NS5-branes, is present in these theories [65]. Since it can be shown that these branes are associated to closed strings on a fundamental level, and not only in the heterotic description, one also has to include them in the type I/II sector of the theory.

The next question to ask is then whether or not one can connect all five formulations of string theory. As it turned out, this question is deeply related to the problem of a strong coupling description for string theory. Since it would go beyond the scope of this introductory chapter, and also since it is not needed in full detail in order to understand the findings of our works, we will simply state the results here.

In fact, it can be argued that every strongly coupled description of string theory is 'S-dual' to another theory that is weakly coupled.[6] To go one by one, the type IIB theory is S-dual to itself [68], while type I and the heterotic theory with gauge group SO(32) are mutually S-dual to each other [55]. The case of type IIA and the heterotic $E_8 \times E_8$ theory is different. These are

---

[6]S-duality was firstly described in the context of compactifications of the heterotic string to four dimensions, where the heterotic string is dual to itself [66,67].

not S-dual to another string description, but to a, up to now, unknown eleven-dimensional theory, called M-theory [56–58,69].

But how is it possible to know the dimension of an otherwise undetermined theory? This can be understood by an argument similar to the one that led to T-duality. In the type IIA theory it is possible to include D0-branes, that is point-like branes. Their mass is always equal and proportional to the inverse string coupling $1/g_s$. Putting a number of n D0-branes on top of each other and going to strong coupling ($g_s \to \infty$) leads then to a continuous spectrum as if a former compact dimension becomes extended again. Hence, one believes that the strong coupling regime of type IIA theory is governed by an eleven-dimensional theory with one compact circle. By going through a duality chain (i.e. by applying several S- and T-dualities), it is possible to show that the heterotic $E_8 \times E_8$ theory is S-dual to the same eleven-dimensional theory, but now with two ten-dimensional boundaries each with one E_8 gauge group on it.

Therefore, all five different string theories, which can be constructed consistently, are connected by dualities and can be seen as different limits of one still unknown theory, that most string theorist nowadays call M-theory. One way to obtain knowledge about M-theory, is to consider the low energy limit of string theory, since there one can shed some light on the part of M-theory that is connected to the type II and heterotic string theories by S-duality. We will therefore discuss this limit next.

Supergravity as low energy limit of string theory

The normal way to obtain the low energy limit of a given theory is to integrate out all fields in the path integral that are heavier then a given mass scale (see e.g. [70–73]). The first question to answer is therefore at which scale one should truncate the spectrum. In string theory there is only one parameter with a mass dimension, the inverse string length $1/l_s$. It is therefore natural to keep only states that are much lighter than $1/l_s$. But since all massive excitations of a string have at least masses of order $1/l_s$ too, it follows that a low energy description of string theory should contain only its massless states.

The next question would then be, which action one should truncate. As one wants to end with a low energy field theory defined in ten dimensions, one cannot use the known world sheet action, but needs a field theoretic description of string theory at high energies. As a matter of principle this is not known.

On the other hand the world sheet theory is perfectly suited to calculate string interactions (see e.g. [30,74–79]). A way to obtain the low energy description is then to deduce it from the S-matrix of string scattering (for the heterotic case see e.g. [80,81]).

Yet another way that could lead to the low energy action is to reconsider the anomalies of the world-sheet theory. In our discussion we claimed that one obtains several EoM-like equations for the states of string theory. Considering only the lowest orders in a string loop expansion should then also give equations that have to be obeyed in the low energy limit (again for the heterotic string see e.g. [82]). The hard task in these approaches is to construct an action that would gives the right scattering amplitudes and equations of motion. We will see in a moment that it is in fact not necessary to deduce the right action in this way.

But before we do so, let us think about what the right expansion parameter for our action should be. Clearly, one should expand the theory in a dimensionless quantity. On the other hand, it should be an expansion that terminates soon if the energy is much smaller then the string mass scale $E \ll 1/l_s$. We therefore conclude that our action should be a power series in $E\,l_s$. Written in this way it becomes obvious that this is a good description, either when the energies are very low, or when l_s is very small.

To lowest order in l_s it is then quite easy to determine the low energy limit of string theory. To this end, one uses the fact that it was mandatory for string theory to be formulated in ten dimensions and to be supersymmetric. As these two properties should not be destroyed in the low energy limit, we are looking for ten-dimensional supersymmetric theories that contain the graviton. The answer to this problem is supergravity in ten dimensions.

Moreover, it turns out that there can be found exactly one ten-dimensional SUGRA for each of the five formulation of string theory, but not more [24, 83–88]. One concludes therefore that indeed supergravity can be used as the low energy limit of string theory. As we will see in the next section there are several advantages in using the supergravity approach to string theory.

1.1.2 Advantages of the supergravity approach

After we have clarified, how string theory and supergravity are connected, one could ask: what is this good for? Why does one consider the low energy limit of a theory of which one has also a description that holds for high energies?

As we already pointed out in the last section, one answer to this question is M-theory, as for M-theory it is not clear what the high energy description is. So, in order to exploit the dualities of string theory fully in a controllable way, it is very useful to consider the low energy regime. To be more precise, it can be shown that the action of type IIA supergravity can be obtained from eleven-dimensional SUGRA by dimensional reduction, i.e. by compactifying on a circle and discarding all massive states. This, together with the fact that SUGRA in eleven dimensions is a unique theory [89], makes it very plausible to view 11d SUGRA as the low energy limit of M-theory.

In the same way one can relate the heterotic $E_8 \times E_8$ supergravity to eleven-dimensional supergravity including two boundaries, confirming also in this sector the claimed dualities [57, 90–92].

Therefore, nearly everything we know about M-theory comes from its low energy description[7], and it is widely customary to use the term M-theory also for its supergravity limit. In this thesis we are mainly interested in the heterotic $E_8 \times E_8$ supergravity and its strong coupling limit, and hence we will reexamine these theories on a more technical level in chapter 2.

Another way to understand the importance of the SUGRA approach to string theory is to remember that string theory has to be formulated in ten dimensions. This is of course in conflict with the fact that it should eventually describe four-dimensional physics. To solve this problem it is inevitable to hide six of the ten dimensions. This is usually done by following the idea of Kaluza-Klein reduction discussed in the last section and by proposing that the six extra dimensions belong to a compact manifold of size so small that it can not be resolved by present experiments.[8] This procedure of compactification is extremely hard to deal with in the original world sheet theory which becomes highly non-linear. On the other hand, compactification is quite easy to handle within supergravity, where it merely corresponds to a special form of the metric.

Furthermore, one can show that the metric of the internal space will affect the physical spectrum that can be observed in four dimensions. Since also the standard model (SM) of particle physics, or more precisely its supersymmetric version (MSSM), should be contained in the low energy description of string theory[9], it is in principle possible to deduce constraints on the compactification from phenomenology and thus find a link from standard model physics to string theory.

One should also note that in the beginning one does not know which compact manifold to take. It is one strength of the supergravity approach to give quite sever constraints on the internal spaces. In fact, a big part of this work is concerned with finding first these constraints and then its solutions.

As a matter of fact, in order to find solutions, one has to think about another problem. Since string theory on the world sheet is formulated as a consistent quantum theory, one should of course also deal with quantized supergravity. However, this is a very difficult thing to do in its own right [99, 100], and it would therefore seem that one has not won a lot in going to the low energy regime. The right viewpoint to this issue is to think about the quantum fields as oscillation around a background field [101]. The background field is the solution to the classical problem, while only the oscillatory part is considered as quantized. In the supergravity approach one considers only the classical solution. This is necessary in order to find the correct and consistent background values of the fields.

In this sense treating sting theory in a supergravity framework is only the first step in a

---

[7]There is also a description of M-theory in terms of matrix-theory [93, 94].

[8]Another way to hide the extra dimensions is to use D-brane models, in which four-dimensional physics is located on a brane [28, 95, 96].

[9]For an introduction to the SM and the MSSM see e.g. [97, 98]. Since string theory is supersymmetric, one can of course only get the MSSM in the low energy limit if one does not invoke SUSY-breaking.

complete quantum mechanical formulation. But the point that should be stressed here is that without having the right classical background it would be even harder – if not impossible – to develop a quantized field theoretic description. This is what makes the supergravity approach on the one hand feasible and on the other hand essential. And it is for this reason that so much work, including our own, has been devoted to this approach.

After having clarified the strength of supergravity in the context of string theory, we also want to point out some of its drawbacks. An obvious downside is that one looses the explicit stringiness of the theory, simply because one is dealing with ordinary field theory. Stringy corrections to the lowest order supergravity action come as power series in the string length l_s and can in principle be inferred from S-matrix elements of the world sheet theory [80–82, 102–104]. Other corrections can be found by anomaly considerations of the supergravity theories and lead to higher order contributions in the action. This is in particular the case for the heterotic theories and its M-theoretic limit [24,90,105]. Therefore, if one wants to calculate string cross sections, i.e. intrinsically stringy quantities, one should definitively not do it within SUGRA, but within the world sheet approach.

A further implication of the SUGRA approach is that one is working at a classical level. Despite of the benefits of this approach, which we listed above, one has to assure that no quantum effects alter the results on a fundamental level. The scale at which one expects that quantum gravity effects to become important is again given by the inverse string length $1/l_s$. Therefore, in flat ten-dimensional space our truncation ansatz $E \ll 1/l_s$ assures that we can treat the problem classically.

But when one starts to compactify the theory, another mass scale $1/R$, determined by the typical compactification radius R, is introduced. In order to guarantee that the classical description still makes sense, one has to demand $E \ll 1/R \ll 1/l_s$, thereby decoupling Kaluza-Klein modes from quantum modes. Hence, there is an intrinsic consistency condition on the supergravity approach, which demands that the length scales of the compactification manifold should not get to small. On the other hand, the seize should not get to big either in order not to be in conflict with four-dimensional physics. This is a generic problem in the supergravity approach, and we discuss one of its solution in the next section.

1.2 The moduli problem in supergravity

After having illuminated the advantages of the supergravity approach in the last section we will focus here on one of its main problems, the issue of four-dimensional massless fields, called moduli. The solution of this problem can be obtained by introducing non-vanishing flux on the compact manifold, i.e. by doing 'flux compactification'.

1.2.1 Calabi-Yau compactification and moduli

We have mentioned above that four-dimensional phenomenology and the internal compact manifold are related. An important example is given by Calabi-Yau (CY) compactifications. It was recognized already some time ago that the amount of supersymmetry preserved during the process of compactification is not predetermined. But a sensible assumption for the energy regime we are interested in is that it should be possible to realize the minimal supersymmetric extension of the standard model, which is an $\mathcal{N}=1$ theory[10], meaning that also the compactified theory should have $\mathcal{N}=1$ SUSY. For the simplest settings this request on four-dimensional physics can be translated into conditions on the internal space, and restricts it to have a vanishing Ricci tensor [27, 106–109]. In compactifications form ten to four dimensions one knows that such spaces exist due to proofs by Calabi and Yau [110, 111]. These spaces are therefore called Calabi-Yau manifolds (for further properties of CY manifolds see e.g. [112–114]). For compactifications of M-theory to four dimensions it is also possible to find explicit examples [115–117].

Having fixed the amount of supersymmetry from the phenomenological viewpoint, one can ask next what kind of particles are to be expected. The answer to this question can be given by explicitly calculating the effective four-dimensional SUSY action and provides the result that one has to expect several massless scalar fields in every case. These scalar fields correspond to flat directions in the potential and are called moduli [8–10]. The disturbing fact about these findings is that such massless bosons should mediate forces in four dimensions that have not been observed [118].

Besides from this conflict with experiment, moduli are also dangerous from the theoretical viewpoint. The reason for this is that the moduli fields correspond to geometrical quantities of the CY manifold [119–123]. As their potential energy is flat, they can take any value they want, leaving the compactification manifold quite undetermined. In particular, there is always one modulus corresponding to the overall size of the manifold. This is very unsatisfactory, since as we have discussed in the previous section the radius should be neither to big nor to small. Therefore, in order to have viable physics in four-dimension, but also to guarantee consistency of the supergravity approach, one is forced to introduce some mechanism that gives a mass to all moduli fields and stabilizes their vacuum expectation values. This in turn will then fix the form and the seize of the internal manifold, respectively. One way to achieve this, which has become very popular in the recent years, is to include non-vanishing background flux into the analysis.

---

[10]This means that SUSY is generated by one four-dimensional Dirac spinor, or equivalently by two Weyl spinors.

1.2.2 Stabilizing moduli with flux

One way to understand the origin of background fluxes in string theory is to look at the fields appearing in the classical supergravity actions. All ten-dimensional SUGRA theories contain besides the metric and a scalar field (the dilaton) various p-form field strengths that are built from $(p-1)$-form potentials analogous to the electromagnetic field strength and its one-form potential.[11] Following the analogy, one can assign fluxes to these p-form field strengths, just as one does in the electromagnetic field. That is, one claims that the p-dimensional integral over a p-form field strength does not vanish for 'magnetic' p-form flux, while one speaks of 'electric' flux when the $(10-p)$-dimensional integral over the field strength is not zero.

Explicit sources for these fluxes are D-branes and NS5-branes. However, just as an electromagnetic wave can propagate without sources, it is also possible to include general p-form flux without sources. Furthermore, an analysis similar to the one that led to the Dirac quantization condition [124] shows that also the more general p-form fluxes have to be quantized [125–127].

In the previous section we claimed that for the simplest settings one has to compactify on Calabi-Yau manifolds, but we did not explain in what sense these settings are simple. The answer is that in the CY case one does set all background fluxes to zero. Due to flux quantization it is impossible to generate flux dynamically, which ensures the consistency of the ansatz.[12] But in principle flux should be included in the analysis. However, it should also not be in conflict with the four-dimensional Poincaré symmetry, i.e. with the known background properties of the observed space-time. This restricts fluxes to be uniform in four-dimensional space-time (when possible), or to be totally localized in the compact part of space.

Non-vanishing flux has in general several implications on the problem of how to compactify the theory in a supersymmetric way. First, SUSY in four dimensions can only be obtained if not all flux components, which are possible in principle, are present. Moreover, fluxes affect heavily the geometry of space-time. For example, the internal space cannot be described as a CY manifold anymore. Since in this case there is no powerful existence theorem as in the CY case, an essential problem is to construct explicitly manifolds that satisfy all conditions required by supersymmetry in order to show that consistent solutions are not absent. Furthermore, fluxes can alter the overall space-time geometry such that four-dimensional space-time obtains an exponential warp-factor, which is an extra scale factor depending on the position in the internal space [129–140].

Considering all these changes, one should not be surprised that fluxes can also be used to stabilize moduli. Keeping the discussion as broad as possible, one can say that fluxes introduce new terms in the effective four-dimensional SUSY action and generate in that way a non-flat

---
[11]Eleven-dimensional supergravity also has a metric and a three-from potential, but no dilaton.
[12]As noticed already in [128] it is possible to change the amount of flux by tunneling processes, which can be neglected in most cases.

potential for some of the moduli [141–149]. It is for this interesting possibility to obtain massive moduli within the classical background approach of supergravity that flux compactification has attracted so much attention in the recent years.[13]

However, there are also other interesting effects that one can study within the field of flux compactification. In particular, as we mentioned before it is possible to find a supersymmetric theory in four dimensions only, when certain conditions on the flux are satisfied. Turning the argument around, one can also ask whether it is possible to break SUSY in a way which resembles low energy SUSY-breaking from the four-dimensional perspective. In the main part of this thesis we will present our results found on this question for $E_8 \times E_8$ supergravity.

One of the main problems which one encounters by dealing with fluxes is that CY manifolds are no solutions any more. Although CY spaces are in itself already a very difficult subject, there are several important mathematical results about them which facilitate the actual work with them [112–114]. Much less is known about the manifolds that appear in the context of flux compactification. But one powerful tool, which can be used in the analysis of fluxes, is the classification of manifolds in terms of G-structures, whose adaption by string theorists helped on the field a lot.

1.3 G-structures and flux compactifications

A first sign for the importance of the language of G-structures for flux compactification is that it includes the Calabi-Yau case (i.e. no flux) as well as the non-CY case. A CY manifold can be specified mathematically in several ways. One is, as we mentioned above, its vanishing Ricci tensor. More important in the context of flux compactification is the notion of manifolds of G-holonomy. This means that a vector which gets transported around any closed loop on the manifold comes back transformed under the group G. Also, the covariant derivative built with the Levi-Civita connection of any tensor or spinor that is invariant under the group G is zero for a G-holonomy manifold (more information about holonomy can be found e.g. in [153–156]). A statement that holds also for compactifications of eleven-dimensional supergravity is then that as long as there is no flux the compactification manifold has to obey G-holonomy in order to give $\mathcal{N} = 1$ SUSY in four dimensions.

The inclusion of fluxes generalizes the G-holonomy manifolds to so-called G-structure manifolds. We will give a short introduction of G-structures in chapter 3, but want to mention the most important points also here. Manifolds with G-structure are not random spaces, but have to satisfy conditions that connect them to the holonomy case. All of them that are relevant for physical compactifications have to allow for (at least) one globally well defined spinor that is invariant

---
[13] Comprehensive reviews which also organize the vast number of publications are [96,150–152].

under G, in order to make $\mathcal{N}=1$ SUSY in four dimensions possible. The covariant derivative of this spinor will not be zero as for G-holonomy mani-folds, but can be understood as representing the torsion of the manifold. The amount of allowed torsion is in turn determined by SUSY and can be used to find a classification of G-structure manifolds and SUSY solutions. For vanishing torsion one gets of course the holonomy case.

The role of flux should then be obvious. Vanishing flux as well as vanishing torsion means that one has to compactify on G-holonomy manifolds. Therefore, one should be able to identify flux and torsion. This is indeed possible and can be understood from the fact that flux not only appears in the action, but also in the supersymmetry conditions of the SUGRA theories. Therefore, SUSY determines the torsion of the manifold precisely via the flux, and hence provides the connection between the two quantities. But by determining the torsion it also determines the geometry of the compact space. The importance of the G-structure approach to compactification is given exactly by this: it makes it possible to clarify the deep relation between geometry and flux, that is otherwise quite obscure.[14] It thereby also makes it possible to classify possible solutions [158–161] and enables one to search for explicit example classes.

As we mentioned briefly at the end of the last section, beside the problem of classifying solutions and stabilizing moduli, also the issue of SUSY-breaking can be approached with flux compactifications. We have argued in the last sections that string theory needs to be supersymmetric and that we also expect the theory to be supersymmetric at intermediate energy regimes, i.e. after compactification. A phenomenological reason for this is that the hierarchy problem of the standard model[15] is more easily solved when SUSY is broken at a scale below the compactification scale (see e.g. [98] and references therein).

From the theoretical viewpoint it is also preferable to have supersymmetry intact in a regime as large as possible. The argument for this are so-called BPS-states. Originally, as a BPS-state a massive supersymmetric state was understood, whose mass saturates a bound given by supersymmetry. [166, 167]. Such states are believed to give information about the strong coupling and high energy regime of a theory, as supersymmetry is not affected by the coupling or the energy scale.

In string theory one can also consider BPS-objects, most notably D-branes and NS5-branes, that preserve a given amount of SUSY. Again, these objects give information about the strong coupling and high energy regime in the low energy supergravity approximation, and therefore

---

[14]An early paper that also recognizes the importance of torsion for compactifications without the use of G-structures is [157]. There, solutions on Ricci-flat non-symmetric coset spaces were constructed with the help of non-vanishing torsion.

[15]The standard model hierarchy problem is concerned with the mass of the higgs field. Naturally the higgs mass should be of the order Λ^2, with Λ the cut-off scale of the SM, while a mass in the range of electroweak symmetry breaking, i.e. the TeV range, is experimentally preferred. This contradiction is softened by low scale SUSY-breaking, due to cancellations of boson and fermion loops. For more details see e.g. [162–165].

tighten the connection between the SUGRA approach and the actual string theory [168–172]. By explicit strong SUSY-breaking in the theory one would loos this information. SUSY-breaking, since it is necessary, should therefore be as mild as possible and adiabatically connected to the supersymmetric case.

A very popular mechanism, which can achieve this, is so-called gaugino condensation. Here, one considers a fermionic condensate which is connected to the SM only by gravity [54, 173–178]. In this way an explicit SUSY-breaking at a high scale in a hidden sector is mediated by gravity to soft SUSY-breaking terms in the observable sector. However, this is a non-perturbative approach, since the formation of a condensate is a strong coupling effect. One could therefore ask whether it is possible to achieve similar effects within a perturbative setting alone.

In fact, this can be achieved by including non-vanishing fluxes. As we have discussed previously it is essential for a supersymmetric vacuum that flux and torsion are in accordance with each other. Stated differently: having the 'wrong' amount of flux will lead to a non-supersymmetric setting (e.g. [133, 142, 143, 179–190]). However, just demanding that SUSY is broken does not imply that one has found a consistent vacuum, i.e. that the equations of motion are satisfied and that one has found a stable minimum of the potential. Furthermore, it is not clear what the phenomenology in four dimensions would be. Therefore, it is mandatory to have an organizing principle if one wants to discuss SUSY-breaking with fluxes.

To this end, the notion of calibrated submanifolds can be used (see e.g. [158, 159, 191–196] and also section 4.3). In fact, the supersymmetry conditions for ten-dimensional supergravity can be interpreted as calibration conditions for various extended objects in four dimensions. Since these will also minimize their energy as long as they are calibrated, one can consider them also as BPS-objects. Starting from this, it was shown in [197] in the context of the type II theories that by allowing the SUSY conditions corresponding to calibrated four-dimensional domain walls to fail a whole family of controllable soft scale SUSY-breaking vacua can be constructed. This type of supersymmetry breaking was therefore called domain wall SUSY-breaking (DWSB).

One of the main goals of our work was to extend this formalism to the case of the heterotic string. To understand the relevance of this extension one should realize that by string duality it is possible to argue that solutions that are found in the type II description of the theory should have dual descriptions in the heterotic string. This was done in [190] for non-supersymmetric solutions of the heterotic string. In our work we confirmed this claim by explicit calculation[16].

After having established DWSB in heterotic supergravity it was also possible to include a gaugino condensate along the lines of [198] and examine its effects on our solutions. Interestingly, SUSY-breaking by gaugino condensation seems to be orthogonal to DWSB in the sense that one cannot cancel mutually the effects of both approaches and get a supersymmetric vacuum.

---

[16]Note that such a check is only possible when both supergravity descriptions are each in the weak coupling limit.

From the perspective of string duality it is quite interesting whether the results we found for the heterotic string can be extended to the strong coupling regime, i.e. to heterotic M-theory. The second part of our work has been devoted to this question.

Important equations, which were already known for the heterotic string had to be rederived first for eleven-dimensional supergravity. Thus, we laid only the foundations for further research in this area, but were not able to answer all question as thoroughly as for heterotic supergravity. However, it was at least possible for us to show that a substantial class of solutions found in ten dimensions can be lifted to heterotic M-theory.

This brings us to the following summary of our work: We are searching for classical solutions of heterotic string theory, i.e. certain background values for the metric, the dilaton, the flux, and the gauge fields, which break supersymmetry in a controllable way. To find these solutions we use the supergravity approach to string theory which is especially well suited for the construction of classical backgrounds. Our search is guided twofold: firstly, by the existence of an organized recipe to construct non-SUSY vacua of the type II string with non-vanishing fluxes, and secondly, by string duality. The inclusion of gaugino condensation and the strong coupling limit connect our results to other fields of research and guarantee in this way its consistency.

1.4 Outline and results

Let us end this introductory chapter by giving a short outline of the thesis and the results achieved in it.

In chapter 2 we briefly review heterotic $E_8 \times E_8$ supergravity in eleven and ten dimensions. We present the actions, the supersymmetry transformations, and how to obtain the ten-dimensional theory as a limit of the eleven-dimensional one. We show that both actions can be rewritten after compactification as four-dimensional integrals over an effective potential which depends only on the internal manifold. The equations of motion derived from these potentials are equivalent to the EoM's obtained from the full action after the compactification ansatz is inserted. This is an important result, which enables us to discuss solely the potential in later chapters.

A main tool of flux compactification in general and of our work in particular is the use of G-structures. We thus give a short introduction to this topic in chapter 3. Our special interest will be on G_2 structures in seven dimensions and $SU(3)$ structures in six and seven dimensions, as these will appear in our compactifications.

Our main results on the weakly coupled heterotic string are introduced in chapter 4, following [1–3]. We start our discussion by rewriting the potential in a BPS-like form, i.e. as a sum of perfect squares of supersymmetry conditions, which means that the potential will be zero and extremized for unbroken SUSY.

In section 4.2 we discuss the implications of DWSB in general. We calculate the dependence of the flux components on the torison of the manifold and compare it to the SUSY case. From the potential we get an additional equation of motion that has to be satisfied in order to obtain a consistent solution. As a further condition from the potential we find that the SUSY-breaking scale should lie well below the compactification scale. Therefore, SUSY-breaking can be viewed as spontaneous in four dimensions. We also calculate the gravitino and gaugino mass and show that they are equal.

After having reviewed the notion of calibration in section 4.3, which gives a geometric interpretation to our model, we restrict our ansatz in section 4.4 in order to solve the residual equation of motion. We show that solutions are given by fiber bundles with two-dimensional fiber and four-dimensional base. From a four-dimensional analysis we obtain the result that this class of solutions correspond to no-scale models in four dimensions, and we confirm that the gravitino and gaugino masses are equal.

In section 4.5 we construct explicit examples that stabilize the dilaton and the radii of the fiber. At the end of chapter 4 we include a gaugino condensate in our considerations and show that it is not possible to cancel the SUSY-breaking effects of DWSB by the effects of gaugino condensation. This shows that the two approaches are orthogonal to each other.

After this extensive discussion of the weak coupling limit, we examine in chapter 5 the question whether our results from ten dimensions can be lifted to the strong coupling limit. To answer this question it is necessary to rewrite the potential of M-theory in a BPS-like form, similar as in the ten-dimensional case. This turns out to be a hard problem and following [4] chapter 5 is almost completely devoted to this issue. The result of our consideration is that in the most general case it is not possible to give the potential of heterotic M-theory in a BPS-like form. We crosscheck this statement by calculating several known limits and can confirm it. But by examining the conditions for BPS-ness more closely, we find that the class of solutions we got in section 4.4 can be lifted and gives a BPS-like potential. This should be seen as a starting point for further research.

In chapter 6 we summarize our results and try to give an outlook on further work. In the appendices we have gathered information about our conventions and several technicalities. In appendix A.6 we also present a few results about flux backgrounds of M-theory which can be reduced to type IIA, that did not appear in the literature yet, but are quite unconnected to the rest of this work.

Chapter 2

Heterotic Supergravity Theories

As was explained in chapter 1 the low energy limit of string theory can be described by various supergravity theories in ten and eleven dimensions. Since in these theories it is easily possible to define the form of the background metric, they serve as the ideal starting ground for flux compactifications, where one demands a spacetime that is split into a four-dimensional external and into a compact internal part. Moreover, it is even possible to see some of the string dualities connecting the different formulations of string theory at the level of its supergravity representatives. As we explained in the introduction we will focus on classical backgrounds. Then, for most calculations it suffices to consider the bosonic part of the SUGRA actions, since all the background values of fermions should be set to zero, in order to preserve Poincaré symmetry. Most important for the results presented in this thesis are eleven-dimensional supergravity with boundaries and heterotic supergravity in ten dimensions. We therefore briefly review the actions and the supersymmetry conditions of these theories in this chapter[1]. We furthermore discuss in general compactifications to four dimensions and show that there exists a four-dimensional effective potential for both theories. An interesting result is that the equations of motion derived from these potentials are equivalent to those derived from the full action. Thus, it is allowed to use the effective potential approach for further investigation.

2.1 Eleven-dimensional supergravity

We start our discussion with eleven-dimensional supergravity. In eleven dimension there is one unique supergravity theory, as was shown in [89]. Furthermore, restricting the field content to particles with a maximal spin of two, one cannot go higher then eleven dimensions if one wants to construct consistent SUGRA theories. This makes eleven-dimensional SUGRA special in its

---

[1] A more thorough examination of all supergravity theories connected to string theory can be found in any textbook on the field, e.g. [5–10].

own right. On the other hand, as discussed in the introduction, it is believed that the various formulations of string theory are different limits of a single eleven-dimensional theory, called M-theory, whose low energy limit is described by eleven-dimensional supergravity (which is most often also called M-theory in this context). In fact, one keystone for this belief is that the strong coupling regimes of type IIA and $E_8 \times E_8$ heterotic string theory can be described by eleven-dimensional SUGRA [56, 57, 90, 92, 199], which makes it possible to interconnect all the known formulations of string theory by dualities. The transition from strong to weak coupling is governed by the size of the extra eleventh dimension. If one takes this dimension to be bounded one arrives at the heterotic theory, while without boundary type IIA SUGRA can be found. We therefore start our discussion with the bulk action, i.e. the action of the theory far from the boundary, and include the effects of the boundary afterwards. Of course, the boundary terms will be absent if one is only interested in the type IIA limit.

2.1.1 The bulk action

The bosonic part of the bulk action is the standard action of eleven-dimensional supergravity [89]

$$S_0 = \frac{1}{2\kappa^2} \int_{X_{11}} \mathrm{dvol}_{11} \left(R^{(11)} - \frac{1}{2} |G_{11}|^2 \right) - \frac{1}{12\kappa^2} \int_{X_{11}} C_{11} \wedge G_{11} \wedge G_{11} , \qquad (2.1.1)$$

where C_{11} is the three-form potential for the four-form flux G_{11}: $\mathrm{d}C_{11} = G_{11}$. The volume form dvol_{11} contains the metric, $\mathrm{dvol}_{11} = \sqrt{-g_{(11)}}\, \mathrm{d}^{11}x$, and the coupling constant κ is related to the eleven-dimensional Planck length l_P by

$$2\kappa^2 = \frac{1}{2\pi} (2\pi l_P)^9 . \qquad (2.1.2)$$

The complete action is invariant under the local supersymmetry transformations[2]

$$\delta E^M{}_M = \bar{\epsilon} \Gamma^M \Psi_M , \qquad (2.1.3)$$
$$\delta (C_{11})_{MNP} = -3\, \bar{\epsilon} \Gamma_{[MN} \Psi_{P]} ,$$
$$\delta \Psi_M = \nabla_M \epsilon + \frac{1}{288} \left(\Gamma_M{}^{NPQR} - 8 \delta^N_M \Gamma^{PQR} \right) (G_{11})_{NPQR}\, \epsilon .$$

Here, $E^M{}_M$ denotes the vielbein associated to the eleven-dimensional metric $g_{(11)}$, Ψ_M is the gravitino, and the SUSY generator ϵ is an eleven-dimensional Majorana spinor. Due to the fact that we want to consider only classical background solutions, preserving Poincaré invariance, all physical spinors (i.e. the gravitino Ψ_M) should vanish. This implies that the bosonic SUSY transformations

---
[2]For a list of our notational conventions, see appendix A.1.

vanish, and that in order to obtain a supersymmetric vacuum also the gravitino variation $\delta\Psi_M$ has to be zero. Therefore, in a supersymmetric vacuum the Killing spinor equation

$$\nabla_M \epsilon + \frac{1}{288}\left(\Gamma_M{}^{NPQR} - 8\delta_M^N \Gamma^{PQR}\right)(G_{11})_{NPQR}\,\epsilon \;=\; 0 \qquad (2.1.4)$$

has to be satisfied.

2.1.2 Effects of the boundary

It was shown by Horava and Witten [57, 90] using anomaly considerations that one can consistently include two ten-dimensional boundaries into eleven-dimensional SUGRA. Each of these boundaries has to carry an E_8 gauge field and contributes an additional piece to the action

$$S_b \;=\; -\frac{1}{8\pi\kappa^2}\left(\frac{\kappa}{4\pi}\right)^{2/3}\sum_{p=1,2}\int_{B_{10,p}}\mathrm{dvol}_{10,p}\left(\mathrm{Tr}\mathcal{F}_p^2 - \frac{1}{2}\mathrm{Tr}(R_p^{(10)})^2\right). \qquad (2.1.5)$$

The two-forms $\mathcal{F}_p$ are the two E_8 field strengths and the trace Tr over the gauge fields is related to the trace in the adjoint representation by $\mathrm{Tr} = \frac{1}{30}\mathrm{tr}_{\mathrm{adj}}$. The trace over the curvature two-forms is defined as

$$\mathrm{Tr}(R_p^{(10)})^2 \;=\; -(R_p^{(10)})^I{}_J \lrcorner (R_p^{(10)})^J{}_I \;=\; \frac{1}{2}(R_p^{(10)})_{IJKL}(R_p^{(10)})^{IJKL}, \qquad (2.1.6)$$

where $R_p^{(10)}$ is the curvature two-form restricted to the p-th boundary. In addition to the Killing spinor equation, which is sufficient to guaranty SUSY for the bulk theory, one has to set also the gaugino variations to zero

$$\delta\chi_p \;=\; -\frac{1}{4}(\Gamma^{IJ}\mathcal{F}_{p\,IJ})\epsilon\,. \qquad (2.1.7)$$

One should note that the boundary terms come with an extra factor of $\kappa^{2/3}$ compared to the bulk action. In principle, there would be further corrections at higher orders of $\kappa^{2/3}$. We will consistently neglect these higher order contributions and restrict ourselves to $\mathcal{O}(\kappa^{2/3})$. Even at this level one finds important alterations to the pure bulk theory.

For our later calculations it is worthwhile to note that one can view Horava-Witten theory from two different perspectives. In the so-called upstairs picture one considers as eleventh dimension the covering space of the orbifold $S_1/\mathbb{Z}_2$ and identifies $x^{11} \sim x^{11} + 2\pi\rho \sim -x^{11}$. Therefore, one obtains two fixed ten-dimensional hyperplanes at $x_1^{11} = 0$ and at $x_2^{11} = \pi\rho$ on which the action S_b

lives. The Bianchi identity for G_{11}, up to $\mathcal{O}(\kappa^{2/3})$, is then given by [57, 92, 199]

$$\frac{1}{4!}(\mathrm{d}G_{11})_{IJKL\,11}\mathrm{d}x^{IJKL} = -\frac{1}{4\pi}\left(\frac{\kappa}{4\pi}\right)^{2/3}\sum_{i=1,2}\delta(x^{11}-x_i^{11})(\mathrm{Tr}\mathcal{F}_i\wedge\mathcal{F}_i - \frac{1}{2}\mathrm{Tr}R_i^{(10)}\wedge R_i^{(10)}), \quad (2.1.8)$$

and the four-form flux G_{11} is fixed to be

$$(G_{11})_{IJKL} = -\frac{1}{8\pi}\left(\frac{\kappa}{4\pi}\right)^{2/3}\left[\varepsilon(x^{11})\,K^{(1)} + \varepsilon(x^{11}+\pi\rho)\,K^{(2)}\right]_{IJKL}, \quad (2.1.9)$$

$$(G_{11})_{IJK\,11} = 3\,\partial_{[I}(C_{11})_{JK]\,11} - \frac{1}{8\pi^2\rho}\left(\frac{\kappa}{4\pi}\right)^{2/3}\left[\omega_3^{(1)}+\omega_3^{(2)}\right]_{IJK}.$$

The periodic step function $\varepsilon(x^{11})$ is defined as

$$\varepsilon(x^{11}) = \mathrm{sign}(x^{11}) - \frac{x^{11}}{\pi\rho} \quad (2.1.10)$$

in the interval $x^{11}\in[-\pi\rho,\,\pi\rho]$ and by periodic continuation outside of this region. We also have used $K^{(p)}$ as a shorthand for the trace terms and introduced the Chern-Simons forms $\omega^{(p)}$, which are defined by

$$K^{(p)} = \mathrm{d}\omega_3^{(p)} = \mathrm{Tr}\mathcal{F}_p\wedge\mathcal{F}_p - \frac{1}{2}\mathrm{Tr}R_p^{(10)}\wedge R_p^{(10)}. \quad (2.1.11)$$

Thus, one sees that the inclusion of the boundary terms lead to a non-trivial Bianchi identity and forces the four-form flux to be non-zero.

In the downstairs picture one takes as eleventh dimension an interval of length $\pi\rho$ with the hyperplanes as boundaries. Then, the action S_0 has to be supplemented with appropriate boundary conditions for the fields [57, 92]. Both of these pictures have their advantages and disadvantages. The upstairs picture is particularly convenient if one performs partial integrations along the eleventh dimension, since the covering space of the orbifold $S_1/\mathbb{Z}_2$ does not have boundaries, but only fixed hyperplanes. On the other hand, in the downstairs picture one really deals with two fixed boundaries, and thus the variations of the fields can be set to zero at the boundaries. We will therefore use the downstairs picture later on when we derive the equations of motion, and the upstairs picture when we perform partial integrals.

A further restriction due to the inclusion of boundaries is that all fields must have definite parity under the orbifold action $\mathbb{Z}_2$. One finds that of the bosonic fields $(g_{(11)})_{IJ}$, $(g_{(11)})_{11\,11}$, and $(C_{11})_{IJ\,11}$ are even and that the rest is odd. The gravitino obeys $\Psi_I(-x_{11}) = \Gamma^{\underline{10}}\Psi_I(x_{11})$ and $\Psi_{11}(-x_{11}) = -\Gamma^{\underline{10}}\Psi_{11}(x_{11})$. Most important for our purposes is that the SUSY generator ϵ becomes chiral on the boundary

$$\Gamma^{\underline{10}}\epsilon(0) = \epsilon(0), \quad (2.1.12)$$

while it had no definite chirality in the pure bulk theory. This restricts the possibilities to decompose ϵ when one compactifies the theory to four dimensions.

In order to obtain heterotic supergravity, one has to dimensionally reduce the bulk action, and take the limit $\rho \to 0$. Then, the flux $(G_{11})_{IJKL}$ becomes zero and the two pieces of the M-theory boundary action will combine to give the first order α'-corrections to heterotic supergravity. We will discuss these issues more thoroughly in section 2.3.

2.2 Heterotic supergravity

Heterotic supergravity can be obtained as the low energy limit of heterotic string theory. As such it is naturally given as an expansion in the Regge-slope parameter $\alpha' = l_s^2/4\pi^2$. But while for the type II theories it is sufficient to work at lowest order in α', for the heterotic theories it is necessary to include the first order corrections in α', in order to make the theories anomaly free. Also, canceling these anomalies demands that heterotic supergravity should include only SO(32) or $E_8 \times E_8$ gauge fields, respectively. Writing the action in the string frame, where one explicitly includes the string coupling $e^{2\Phi}$ in the Einstein-Hilbert term, the bosonic sector of ten-dimensional $\mathcal{N} = 1$ heterotic supergravity[3], up to order $\mathcal{O}(\alpha')$, is given by [200]

$$S = \frac{1}{2\kappa_{10}^2} \int d^{10}x \sqrt{-\det g}\, e^{-2\Phi} \left[\mathcal{R}_{X_{10}} + 4(d\Phi)^2 - \frac{1}{2}H^2 + \frac{\alpha'}{4}(\mathrm{Tr}\, R_+^2 - \mathrm{Tr}\, F^2) \right], \quad (2.2.1)$$

with $2\kappa_{10}^2 = (2\pi)^7 \alpha'^4$. Here, Φ is the dilaton and $\mathcal{R}_{X_{10}}$ is the Ricci scalar of the full ten-dimensional space. The two-form F is the SO(32) or $E_8 \times E_8$ field strength, and the trace Tr is related to the trace in the adjoint representation $\mathrm{tr}_{\mathrm{adj}}$ by $\mathrm{Tr} = \frac{1}{30} \mathrm{tr}_{\mathrm{adj}}$ as in the action (2.1.5). Since we are interested in the connection between M-theory and heterotic supergravity in the end, we will only consider the gauge group $E_8 \times E_8$ in what follows. The curvature two-forms $R^L_{\pm J} = \frac{1}{2} R^L_{\pm JKL} dx^K \wedge dx^L$ are constructed using the connections

$$\omega^L_{\pm JK} = \omega^L_{JK} \pm \frac{1}{2} H^L_{JK}, \quad (2.2.2)$$

with ω^L_{JK} denoting the ordinary Levi-Civita spin connection. Their traces are defined as

$$\mathrm{Tr}\, R_\pm^2 = -R^L_{\pm J} \cdot R^J_{\pm L} = \frac{1}{2} R_{\pm IJKL} R_\pm^{IJKL}. \quad (2.2.3)$$

---

[3] Note that so far we are restricting our discussion to purely bosonic heterotic configurations. However, one may add non-trivial fermionic condensates to the bosonic background, a possibility that we will consider in section 4.6.

Finally, H stands for the Neveu-Schwarz (NS) three-form flux and is given by

$$H = dB + \frac{\alpha'}{4} \omega_{\text{het}} . \qquad (2.2.4)$$

Here, B is the Neveu-Schwarz two-form potential and ω_{het} a Chern-Simons form defined by $d\omega_{\text{het}} = (\text{Tr } R_+ \wedge R_+ - \text{Tr } F \wedge F)$. The Bianchi identity (BI) of H reads then

$$dH = \frac{\alpha'}{4} \left(\text{Tr } R_+ \wedge R_+ - \text{Tr } F \wedge F \right) . \qquad (2.2.5)$$

with $\text{Tr } R_+ \wedge R_+ = -R_{+\underline{J}}^{\underline{I}} \wedge R_{+\underline{I}}^{\underline{J}}$.

One should note here that the term $\text{Tr } R_+^2$ besides being of order α' is also a higher derivative term and might be neglected therefore. However, it is necessary to keep this term, in order to provide an anomaly free theory. An ambiguity arising here is which connection should be used in constructing the involved curvature forms, as the connection is not essential for anomaly cancellation [201, 202]. However, using the connection ω_+ is of advantage in analyzing flux compactifications of heterotic string theory (see e.g. [137, 190, 200]), and we will therefore also use the torsionful connection ω_+ in our discussion.

In order to construct supersymmetric bosonic backgrounds one needs to make sure that the supersymmetry variations of the gravitino Ψ_M, dilatino λ, and gaugino χ vanish. At leading order in α' these are given by[4]

$$\delta \Psi_I = \left(\nabla_I - \frac{1}{4} \slashed{H}_I \right) \epsilon = \nabla_I^- \epsilon = 0 , \qquad (2.2.6a)$$

$$\delta \lambda = \left(\slashed{\partial} \Phi - \frac{1}{2} \slashed{H} \right) \epsilon = 0 , \qquad (2.2.6b)$$

$$\delta \chi = \frac{1}{2} \slashed{F} \epsilon = 0 , \qquad (2.2.6c)$$

where $H_I = \iota_I H$. Note that these supersymmetry transformations are not corrected at order α' [200]. We will see in the next section how to obtain this action from eleven-dimensional supergravity with two boundaries.

2.3 Dimensional reduction of M-theory

In order to obtain the heterotic (or type IIA) theory from M-theory, one has to perform a dimensional reduction. This means that one integrates out one dimension whose length is set to zero

---
[4]We denote the SUSY generator, the gravitino, and the gaugino with the same symbols as in the eleven-dimensional theory in order to keep the notation simple. Moreover, the eleven-dimensional fields get identified with the ten-dimensional ones after dimensional reduction, which justifies this notation.

afterwards. By this procedure one only keeps the zero modes of all fields, and therefore one does not have to care about massive modes that could appear in compactifications to non-zero volumes. In order to obtain a perfect match between the reduced theory and the ten-dimensional action and its SUSY transformations, one also has to reparametrize the metric and the gravitino. In addition, including the boundary, all fields that are odd under the $\mathbb{Z}_2$-action on the extra dimension $S_1/\mathbb{Z}_2$ will be projected out, once the length of the extra dimension is set to zero. We start our discussion with the reduction of the bulk and boundary actions (2.1.1) and (2.1.5). Afterwards, we will discuss the SUSY variations.

Reduction of the bulk action

The most general ansatz for the reduction to ten dimensions would be to start with a metric of the form

$$\mathrm{d}s_{11}^2 = \mathrm{d}s_{10}^2(x_I, x_{11}) + \left(U_1(x_I, x_{11})\mathrm{d}x^I + U_2(x_I, x_{11})\mathrm{d}x^{11}\right)^2 . \tag{2.3.1}$$

However, since we are only interested in the zero modes of all fields, we can drop the dependence on x_{11}. In order to get to the desired result immediately (see e.g. [9, 10]), one also introduces factors of $e^{\Phi/3}$ including the ten-dimensional dilaton Φ, i.e. we use the ansatz

$$\mathrm{d}s_{11}^2 = e^{-\frac{2}{3}\Phi}\mathrm{d}s_{10}^2 + e^{\frac{4}{3}\Phi}(\mathrm{d}x^{11} + X_1)^2 , \tag{2.3.2}$$

where X_1 is a one-form with legs only in the ten large dimensions. In a similar way, one has to decompose the three-form potential C_{11} and the four-form flux G_{11} into parts containing one leg along the extra dimensions and parts that live only in the ten large dimensions. Putting all these pieces together would produce the action of type IIA supergravity and is described in every textbook on the subject (see e.g. [9, 10]). But since we are not so much interested into the type IIA theory, but into heterotic SUGRA we will not follow this route, but immediately take the influence of the boundary into account.

The zero modes of the eleven-dimensional fields will provide the ten-dimensional fields. In order to obtain these modes one has to average all fields over the eleventh dimension (in the upstairs picture) [92, 199]

$$Y^{(0)} = \frac{1}{2\pi\rho}\int_{-\pi\rho}^{\pi\rho} Y^{(11)} \,\mathrm{d}x^{11} . \tag{2.3.3}$$

From this one immediately sees that all fields that are odd under the $\mathbb{Z}_2$-action on the extra dimension can be set to zero in the reduced action. First, we observe that $(g_{(11)})_{I\,11}$ is odd, and will hence vanish. Therefore, the one-form potential X_1 will be zero. Also, the three-form potential and the four-form flux with only ten-dimensional indices are odd and will give no contribution in ten dimensions. Thus, the Chern-Simons term in (2.1.1) will vanish after dimensional reduction, and the flux is reduced to a three form

$$(G_{11}^{(0)})_{IJKL} = \frac{1}{2\pi\rho} \int_{-\pi\rho}^{\pi\rho} (G_{11})_{IJKL} \, dx^{11} = 0 \,, \tag{2.3.4}$$

$$(G_{11}^{(0)})_{IJK\,11} = H_{IJK} = \frac{1}{2\pi\rho} \int_{-\pi\rho}^{\pi\rho} (G_{11})_{IJK\,11} \, dx^{11}$$

$$= 3\, \partial_{[I}(C_{11})_{JK]\,11} - \frac{1}{8\pi^2\rho}\left(\frac{\kappa}{4\pi}\right)^{2/3}\left(\omega^{(1)} + \omega^{(2)}\right)_{IJK}.$$

Since in the limit $\rho \to 0$ the two boundaries become coincident, one can combine the two Chern-Simons forms $\omega^{(p)}$ into the (negative) heterotic Chern-Simons form ω_{het}. Furthermore, the coupling constant κ and the radius ρ are related to the string coupling g_s and α' by

$$\kappa^2 = \frac{1}{2}(2\pi)^8 g_s^3 (\alpha')^{9/2} \,, \tag{2.3.5}$$

$$\rho = g_s (\alpha')^{1/2} \,.$$

Using these relations and defining $C_{IJ\,11} = B_{IJ}$ yields exactly the heterotic NS flux

$$H = dB + \frac{\alpha'}{4}\omega_{\text{het}} \,. \tag{2.3.6}$$

In an analogous way one can show that the zero mode contribution of the elven-dimensional Bianchi identity (2.1.8) gives the heterotic BI (2.2.5).

Taking into account the factors of e^Φ in the metric decomposition (2.3.2), it is then an easy task to see that the bulk action of Horava-Witten theory reduces to the lowest order contribution of the heterotic SUGRA action. The $\mathcal{O}(\alpha')$ contributions will descend from the boundary action of heterotic M-theory, as we will show in the next paragraph.

Reduction of the boundary action

In order to make contact to the $\mathcal{O}(\alpha')$ contributions of (2.2.1), one has to deal with two subtleties. Firstly, as was discussed in section 2.2 in heterotic flux compactifications it is useful to use a torsionful connection ω_+, in order to construct the $\text{Tr}R^2$ term (see e.g. [137, 190, 200]), while in M-theory one uses the Levi-Civita connection. However, as the connection is not essential for anomaly cancellation [201, 202], it is possible to use the torsionful connection also in heterotic M-theory. Thus, one should replace R with $R_+ = R(\omega_+)$ in the boundary action (2.1.5) and the

Bianchi identity (2.1.8) if one wants to make contact to heterotic flux compactifications. Secondly, in order to put the action into string frame one has to use the metric (2.3.2). This will lead to additional contributions coming from the $\text{Tr} R_+^2$ term. Since these terms are of fourth order in derivatives and are not necessary to ensure anomaly cancellation, we will consider them as higher order contributions and neglect them in our discussion [92, 199].

But then the reduction becomes quite trivial. In the limit $\rho \to 0$ the two boundaries get on top of each other, which means that the range of integration for the two summands in (2.1.5) becomes equal. Therefore, the two curvature contributions will add up and the two E_8 gauge terms will sum to give one $E_8 \times E_8$ term. Using (2.3.5) and $\kappa^2 = 2\pi\rho \kappa_{10}^2$ it follows that one encounters also the right prefactor of $\frac{\alpha'}{8\kappa_{10}^2}$ in front of the integral. Thus, we have shown that indeed the action of Horava-Witten theory reduces correctly to the action of heterotic supergravity as it should be. The last step to proof the equivalence is now to determine how the heterotic SUSY variations emerge from the eleven-dimensional ones.

Reduction of supersymmetry variations

We will show in this section how to reduce the supersymmetry transformations of heterotic M-theory (2.1.4) and (2.1.7) to those of heterotic supergravity (2.2.6). In our discussion we will follow closely [10]. We will focus first on the SUSY variation of the elven-dimensional gravitino $\delta\Psi_M$ and show how one can obtain the variations of the ten-dimensional gravitino (2.2.6a) and of the dilatino (2.2.6b). The result will be that the dilatino is essentially the eleventh component of Ψ_M, while the ten-dimensional gravitino is a shifted version of its eleven-dimensional counterpart. We will neglect the boundary in the beginning. This will give us the SUSY transformations of type IIA supergravity as an intermediate result. Only in the end we will take into account the boundary effects, which will result in projecting out various components of the flux.

We start by rewriting the flux part of (2.1.4), using tangent space indices

$$\delta\Psi_{\underline{M}} = \left(\partial_{\underline{M}} + \frac{1}{4}\omega^{(11)}_{\underline{MNP}}\Gamma^{\underline{NP}} + \frac{1}{24}\left(3\mathcal{G}_{11}\Gamma_{\underline{M}} - \Gamma_{\underline{M}}\mathcal{G}_{11}\right)\right)\epsilon \,. \quad (2.3.7)$$

In order to be able to reduce this to ten dimensions, we have to know how the eleven-dimensional spin connection[5] $\omega^{(11)}_{\underline{MNP}}$ behaves under the reduction (2.3.2). After some calculation one finds

$$\omega^{(11)}_{\underline{IMN}}\Gamma^{\underline{MN}} = e^{\Phi/3}\left(\omega^{(10)}_{\underline{IJK}}\Gamma^{\underline{JK}} - \frac{2}{3}\Gamma_{\underline{I}}{}^{\underline{J}}\partial_{\underline{J}}\Phi\right) + e^{4\Phi/3}(\mathrm{d}X_1)_{\underline{IJ}}\Gamma^{\underline{J}}\Gamma_{\underline{10}} \,, \quad (2.3.8)$$

$$\omega^{(11)}_{\underline{10}\,\underline{MN}}\Gamma^{\underline{MN}} = -\frac{1}{2}e^{4\Phi/3}(\mathrm{d}X_1)_{\underline{IJ}}\Gamma^{\underline{IJ}} - \frac{4}{3}e^{\Phi/3}\Gamma^{\underline{I}}\Gamma_{\underline{10}}\partial_{\underline{I}}\Phi \,, \quad (2.3.9)$$

---
[5] Any spin connection is defined as $\omega_{MNP} = \frac{1}{2}(-\Omega_{MNP} + \Omega_{NPM} + \Omega_{PMN})$, with $\Omega_{MN}{}^P = 2\partial_{[N}E^P_{M]}$ and E^N_M the corresponding vielbein.

where all contractions on the right hand side are done with the purely ten-dimensional metric. As a second step we define the heterotic dilatino and the heterotic gravitino to be

$$\lambda = -3\, e^{-\Phi/6}\, \Psi_{\underline{10}}\,, \qquad (2.3.10)$$

$$\Psi_{\underline{I}}^{(10)} = e^{-\Phi/6}\left(\Psi_{\underline{I}} + \frac{1}{2}\Gamma_{\underline{I}}\Gamma_{\underline{10}}\Psi_{\underline{10}}\right). \qquad (2.3.11)$$

Defining also $\epsilon^{(10)} = e^{\Phi/6}\epsilon$, one finds that the variations of these fields are given by

$$\delta\lambda = \Big(\Gamma^I \partial_I \Phi \Gamma_{\underline{10}} - \frac{1}{12}(G_{11})_{IJK\,\underline{11}}\Gamma^{IJK} + \frac{3}{8} e^{\Phi}(\mathrm{d}X_1)_{IJ}\Gamma^{IJ} \qquad (2.3.12)$$
$$- \frac{1}{96} e^{\Phi}(G_{11})_{IJKL}\Gamma^{IJKL}\Gamma_{\underline{10}}\Big)\epsilon^{(10)}\,,$$

$$\delta\Psi_I^{(10)} = \Big(\nabla_I - \frac{1}{8}(G_{11})_{IJK\,\underline{11}}\Gamma^{JK}\Gamma_{\underline{10}} - \frac{1}{8} e^{\Phi}(\mathrm{d}X_1)_{JK}\Gamma_I^{JK}\Gamma_{\underline{10}} \qquad (2.3.13)$$
$$+ \frac{1}{192} e^{\Phi}(G_{11})_{J_1 J_2 J_3 J_4}\Gamma^{J_1 J_2 J_3 J_4}\Gamma_I\Big)\epsilon^{(10)}\,.$$

These are the SUSY transformations corresponding to type IIA theory. To obtain the heterotic equations one simply has to take into account that the purely ten-dimensional components of G_{11} and the field X_1 are projected out by the boundary conditions, as discussed in the previous section. Furthermore, the chirality condition on the boundary (2.1.12) will hold in general in the reduced theory, which means that the SUSY generator $\epsilon^{(10)}$ will have definite chirality as required for the heterotic theory. Finally, defining $(G_{11})_{IJK\,\underline{11}} = H_{IJK}$, one obtains

$$\delta\lambda = \Big(\Gamma^I \partial_I \Phi \Gamma_{\underline{10}} - \frac{1}{12} H_{IJK}\Gamma^{IJK}\Big)\epsilon^{(10)} \;=\; \Big(\displaystyle{\not}\partial\Phi - \frac{1}{2}\displaystyle{\not}H\Big)\epsilon^{(10)}\,, \qquad (2.3.14)$$

$$\delta\Psi_I^{(10)} = \Big(\nabla_I - \frac{1}{8} H_{IJK}\Gamma^{JK}\Big)\epsilon^{(10)} \;=\; \Big(\nabla_I - \frac{1}{4}\displaystyle{\not}H_I\Big)\epsilon^{(10)}\,. \qquad (2.3.15)$$

These are exactly the SUSY equations for the heterotic dilatino and gravitino, respectively.

Comparing the SUSY transformations of the gauginos in eleven dimensions (2.1.7) to their corresponding transformation in ten dimensions (2.2.6c), one immediately sees that they agree after the field redefinition $\chi_1 + \chi_2 = -e^{\Phi/6}\chi$ and $\mathcal{F}_1 + \mathcal{F}_2 = F$.

It is therefore obvious that not only the action, but also the supersymmetry variation of heterotic supergravity can be obtained from the eleven-dimensional heterotic M-theory.

After this more general introduction, we will discuss common properties of both theories under compactifications to four dimensions in the last section of this chapter.

2.4 Scalar potential and equations of motion

In the last three sections we have discussed heterotic M-theory and heterotic supergravity in general. We also explained how to obtain one theory from the other by dimensional reduction. In this section we will show (following [1–4, 197]) that by compactifying both theories to four dimensions it is possible to rewrite the action as a four-dimensional integral over an effective scalar potential. For the eleven-dimensional as well as for the ten-dimensional theory the equations of motion obtained from this potential are equivalent to those deduced from the full action. One is therefore allowed to use the effective potential for the investigation of four-dimensional physics in general. In particular, we will use the results of this section to discuss the existence of consistent non-supersymmetric vacua of heterotic supergravity in chapter 4 and their lift to M-theory in chapter 5.

2.4.1 Heterotic M-theory

We will start our analysis again with Horava-Witten theory. Here, the scalar potential will have two contributions, one from the bulk and one from the boundary. In the boundary part we will perform an extra Weyl rescaling of the metric, in order to obtain better comparability with the heterotic theory.

The scalar potential

Since we want our compactification ansatz to maintain four-dimensional Poincaré invariance, the metric and the four-form flux should be decomposed as

$$\mathrm{d}s_{11}^2 = e^{2A}\mathrm{d}\hat{s}_4^2 + \mathrm{d}s_7^2 \,, \tag{2.4.1}$$

$$C_{11} = \frac{1}{3!}(C^{(4)})_{\mu\nu\sigma}\mathrm{d}x^{\mu\nu\sigma} + \frac{1}{3!}C_{mnp}\mathrm{d}x^{mnp} \,, \tag{2.4.2}$$

$$G_{11} = \tilde{\mu}\,\hat{\mathrm{dvol}}_4 + \frac{1}{4!}G_{mnpq}\mathrm{d}x^{mnpq} \,. \tag{2.4.3}$$

Here, A is a warp factor, $\mathrm{d}\hat{s}_4^2$ is the metric of a maximally symmetric four-dimensional space-time[6], and $\mathrm{d}s_7^2$ is the metric of a seven-dimensional compact manifold M. The volume element of the maximally symmetric four-dimensional space is denoted by $\hat{\mathrm{dvol}}_4$, $\tilde{\mu}$ is a real constant, and G is the four-form flux restricted to the seven internal dimensions. Inserting (2.4.1) and (2.4.3) into

---
[6]A maximally symmetric space-time has a constant curvature scalar R. The possible spaces are therefore Minkowski space ($R = 0$), anti de Sitter space ($R < 0$) and de Sitter space ($R > 0$).

the bulk action (2.1.1) gives the bulk part V_0 of the scalar potential

$$\begin{aligned}S_0 &= \frac{1}{2\kappa^2}\int_{X_4}\hat{\mathrm{dvol}}_4 \int_M \mathrm{dvol}_7 \left[e^{2A}\hat{R}^{(4)} + e^{4A}(R - 8\nabla^2 A - 20\,\mathrm{d}A^2 - \frac{1}{2}|G|^2 \right.\\ &\quad \left. - \frac{1}{2}\mu^2 - \mu\, C\lrcorner * G) \right] \\ &= \frac{1}{2\kappa^2}\int_{X_4}\hat{\mathrm{dvol}}_4 \int_M \mathrm{dvol}_7 \left[e^{2A}\hat{R}^{(4)} - \mathcal{V} \right] = -\int_{X_4}\hat{\mathrm{dvol}}_4\, V_0\,. \end{aligned} \qquad (2.4.4)$$

Here, R denotes the Ricci scalar of M, and we introduced $\mu = e^{-4A}\tilde{\mu}$. Since the external space is maximally symmetric, the unwarped four-dimensional curvature scalar $\hat{R}^{(4)}$ is constant and all other fields depend only on the seven-dimensional internal space M.

On the boundary the eleven-dimensional metric splits into a ten- and a one-dimensional piece

$$\mathrm{d}s_{11}^2 = \mathrm{d}s_{10}^2 + v\otimes v\,, \qquad (2.4.5)$$

with $\mathrm{d}s_{10}^2$ the metric on the boundary and v a one-form perpendicular to the boundary. In order to get results that can be used easily if one wants to compare to the weakly coupled heterotic theory, we perform a Weyl rescaling of the metric similar to (2.3.2) $g_{10} \to g'_{10} = e^{-\sigma}g_{10}$.[7] Note that we do not relate the rescaling to the dilaton Φ but keep it unspecified for the moment by introducing the field σ. This rescaling will introduce terms with four derivatives coming from the $\mathrm{Tr}R^2$ terms in the boundary action, which we neglect in our analysis, since they are of higher order. The metric g'_{10} is then compactified according to

$$\mathrm{d}(s'_{10})^2 = e^{2A'}\mathrm{d}\hat{s}_4^2 + \mathrm{d}(s'_6)^2\,, \qquad (2.4.6)$$

where A' is a shifted warp factor $A' = A + \frac{1}{2}\sigma$. Taking this into account, the boundary action can be written as

$$\begin{aligned}S_b &= -\frac{1}{8\pi\kappa^2}\left(\frac{\kappa}{4\pi}\right)^{2/3}\sum_{p=1,2}\int_{X_4}\hat{\mathrm{dvol}}_4 \int_{B_{6,p}} \mathrm{dvol}'_{6,p}\, e^{4A'-3\sigma}\left\{\left(\mathrm{Tr}(\mathcal{F}_p^{(6)})^2 - \frac{1}{2}\mathrm{Tr}(R_{p+}^{(6)'})^2\right)\right.\\ &\quad \left. -\frac{1}{24}\left|e^{-2A'}\hat{R}^{(4)} - 12|\mathrm{d}A'|^2\right|^2 - 4\,e^{-2A'}(\nabla_i\nabla_j e^{A'})(\nabla^i\nabla^j e^{A'}) - 2\left|\mathrm{d}A'\lrcorner H\right|^2\right\}. \end{aligned} \qquad (2.4.7)$$

Here, we used that the E_8 gauge fields are confined to the internal space $\mathcal{F} = \mathcal{F}^{(6)}$. The appearing three form field $H_{ijk} = G_{ijk\,11}$ is used to construct the torsionful curvature tensor R_+ in the same way as the Neveu-Schwarz three form is used as torsion in the heterotic string. Note that we kept

---
[7]At the same time one should rescale the $(11,11)$-component of the metric by $e^{2\sigma}$. However, this cannot be seen at the boundary.

four derivative terms in this expression although we discarded them in our previous discussion. We do this only in order to compare to the results found in [1–3] for the heterotic string that will be presented later.

The action can thus be written as a four-dimensional integral over a scalar potential

$$S = -\int_{X_4} \hat{\text{dvol}}_4 \, V = -\int_{X_4} \hat{\text{dvol}}_4 \, (V_0 + V_b), \tag{2.4.8}$$

combining the contribution from the bulk V_0 and from the boundary V_b, respectively, which are given by

$$V_0 = \frac{1}{2\kappa^2} \int_M \text{dvol}_7 \left\{ -e^{2A} \hat{R}^{(4)} - e^{4A} \left(R - 8\nabla^2 A - 20\,\text{d}A^2 - \frac{1}{2}|G|^2 - \frac{1}{2}\mu^2 - \mu\, C \lrcorner * G \right) \right\}, \tag{2.4.9a}$$

$$V_b = \frac{1}{8\pi\kappa^2}\left(\frac{\kappa}{4\pi}\right)^{2/3} \sum_{p=1,2} \int_{B_{6,p}} \text{dvol}'_{6,p}\, e^{4A'-3\sigma} \left\{ \left(\text{Tr}(\mathcal{F}_p^{(6)})^2 - \frac{1}{2}\text{Tr}(R_{p+}^{(6)'})^2 \right) \right.$$
$$\left. - \frac{1}{24}\left| e^{-2A'}\hat{R}^{(4)} - 12|\text{d}A'|^2\right|^2 - 4\,e^{-2A'}\,(\nabla_i\nabla_j e^{A'})(\nabla^i\nabla^j e^{A'}) - 2\left|\text{d}A' \lrcorner H\right|^2 \right\}. \tag{2.4.9b}$$

Equations of motion

We will now show that the equations of motion derived from (2.4.8) are consistent with the full eleven-dimensional equations coming from (2.1.1) and (2.1.5). In order to proof this, we will make use of the downstairs picture. Setting the variations of the fields at the boundaries to zero, as is usual, we are left with the bulk action plus boundary conditions [92]. Since these boundary conditions will not change in going from eleven to four dimensions, we only have to consider the bulk part of (2.4.8).

Varying (2.1.1) with respect to the metric $g^{(11)}$ and the three form potential C_{11}, one obtains

$$\delta_{g^{(11)}}: \quad R_{MN}^{(11)} - \frac{1}{2}g_{MN}^{(11)}\left[R^{(11)} - \frac{1}{2}|G_{11}|^2\right] - \frac{1}{2}(\iota_M G_{11})\lrcorner(\iota_N G_{11}) = 0, \tag{2.4.10}$$

$$\delta_{C_{11}}: \quad \text{d}*_{11} G_{11} + \frac{1}{2}G_{11} \wedge G_{11} = 0. \tag{2.4.11}$$

Restricting (2.4.10) to internal coordinates $MN = mn$ and inserting (2.4.3) leads to

$$R_{mn} - 4\,e^{-A}\nabla_m\nabla_n e^A - \frac{1}{2}\iota_m G \lrcorner \iota_n G \qquad (2.4.12)$$
$$-\frac{1}{2}g_{mn}\Big[e^{-2A}\hat{R}^{(4)} + R - 8\,\nabla^2 A - 20\,\mathrm{d}A^2 - \frac{1}{2}\mu - \frac{1}{2}|G|^2\Big] = 0\,.$$

Taking the trace of (2.4.10) over its external indices one finds

$$6\,\nabla^2 e^{2A} = \hat{R}^{(4)} + 2\,e^{2A}\Big(R + \frac{1}{2}\mu - \frac{1}{2}|G|^2\Big)\,. \qquad (2.4.13)$$

Furthermore, inserting (2.4.3) into the equation of motion for C_{11} (2.4.11) gives

$$\mathrm{d} * G = -\mu G\,. \qquad (2.4.14)$$

These are exactly the EoM's that one obtains if one varies the bulk part of (2.4.8) with respect to the internal metric g^{mn}, the warp factor A, and the flux G, respectively. Thus, by inserting our compactification ansatz into the equations of motion of the eleven-dimensional theory, we obtain the same results as if we vary (2.4.8). It is therefore sufficient to work with the effective potential instead of the full eleven-dimensional action. This will simplify our analysis during the rest of this work.

2.4.2 Scalar potential of heterotic supergravity

We now repeat the same steps as before for heterotic supergravity. We will see again that an effective potential is sufficient in order to analyze the resulting four-dimensional physics. In addition, in ten dimensions it is possible to obtain quite severe restrictions on the warp factor and the four-dimensional cosmological constant from the equations of motion.

Four-dimensional compactifications and effective potential

We restrict our attention again to compactifications to four dimensions and assume that the ten-dimensional spacetime X_{10} splits into a six-dimensional compact manifold M_h and a maximally symmetric space X_4 with cosmological constant Λ

$$\mathrm{d}s^2_{X_{10}} = e^{2A_h}\mathrm{d}s^2_{X_4} + \mathrm{d}s^2_{M_h}\,. \qquad (2.4.15)$$

Moreover, in order to keep our construction as general as possible, we allow for non-trivial warping A_h, dilaton Φ, and fluxes H and F such that the 4d maximal symmetry is not broken. As a result, the ten-dimensional BI (2.2.5) keeps the same form, but with R_+ being just the curvature of the internal torsionful connection $\omega^i_{+jl} = \omega^i_{jl} + \frac{1}{2}H^i_{jl}$. In the following, $R_\pm$ will denote this internal torsionful curvature.

Inserting our ansatz into the heterotic action (2.2.1), and setting $S = -\int_{X_4} \mathrm{d}^4 x \, V_h$, the resulting four-dimensional potential is given by

$$\begin{aligned}
V_h &= -\frac{1}{2\kappa_{10}^2} \int_{M_h} \mathrm{dvol}_{M_h} \, e^{4A_h - 2\Phi} \Big\{ e^{-2A_h} R_{X_4} + \mathcal{R} - \frac{1}{2}H^2 - 4(\mathrm{d}\Phi)^2 - 8\nabla^2 A_h \\
&\quad -20(\mathrm{d}A_h)^2 + \frac{\alpha'}{4}(\mathrm{Tr}\, R_+^2 - \mathrm{Tr}\, F^2) \\
&\quad + \alpha' \Big[2e^{-2A_h}(\nabla^i \nabla^j e^{A_h})(\nabla_i \nabla_j e^{A_h}) + |\mathrm{d}A_h \lrcorner H|^2 + 3 \Big| \frac{1}{12} e^{-2A_h} R_{X_4} - \mathrm{d}A_h \lrcorner \mathrm{d}A_h)^2 \Big|^2 \Big] \Big\},
\end{aligned} \quad (2.4.16)$$

where $\mathcal{R}$ is the scalar curvature constructed using the internal six-dimensional metric, and the last line of (2.4.16) arises from the fact that now R_+ denotes just the internal curvature.

Equations of motion

In order to derive the equations of motion from the action (2.2.1), a little care is needed, due to the implicit dependence of H on other elementary fields through the BI (2.2.5), and due to of the presence of the order α' curvature corrections. However, this complication is simplified by a lemma stating that the variation of the action with respect to the torsionful curvature connection ω_+ is proportional to the leading order equations of motion [200, 203]. In an approach that solves the EoM's order by order in α' one can set these contributions to zero. It is therefore sufficient to vary only the explicitly appearing fields if one wants to derive the equations of motion. The EoM's read then

$$\begin{aligned}
\delta_g: \quad & (\mathcal{R}_{X_{10}})_{IJ} + 2\nabla_I \nabla_J \Phi - \frac{1}{2}\iota_I H \cdot \iota_J H + \frac{\alpha'}{4}\big[\mathrm{Tr}(\iota_I R_+ \cdot \iota_J R_+) - \mathrm{Tr}(\iota_I F \cdot \iota_J F) \big] \\
& - \frac{1}{2} g_{IJ} \Big(\mathcal{R}_{X_{10}} - 4|\mathrm{d}\Phi|^2 + 4\nabla^2 \Phi - \frac{1}{2}|H|^2 + \frac{\alpha'}{4}\big[\mathrm{Tr}\, R_+^2 - \mathrm{Tr}\, F^2 \big] \Big) = 0 \,, \quad (2.4.17) \\
\delta_\Phi: \quad & \mathcal{R}_{X_{10}} - 4|\mathrm{d}\Phi|^2 + 4\nabla^2 \Phi - \frac{1}{2}|H|^2 + \frac{\alpha'}{4}\big[\mathrm{Tr}\, R_+^2 - \mathrm{Tr}\, F^2 \big] = 0 \,, \quad (2.4.18) \\
\delta_B: \quad & \mathrm{d}(e^{-2\Phi} *_{10} H) = 0 \,, \quad (2.4.19) \\
\delta_\mathcal{A}: \quad & e^{2\Phi} \mathrm{d}(e^{-2\Phi} *_{10} F) + \mathcal{A} \wedge *_{10} F - *_{10} F \wedge \mathcal{A} - F \wedge *_{10} \mathrm{d}B = 0 \,, \quad (2.4.20)
\end{aligned}$$

with $\mathcal{A}$ being the gauge potential of F. The dilaton EoM can be used to simplify the Einstein equation. The resulting EoM is called 'modified Einstein equation' and reads

$$(\mathcal{R}_{X_{10}})_{IJ} + 2\nabla_I \nabla_J \Phi - \frac{1}{2}\iota_I H \cdot \iota_J H + \frac{\alpha'}{4}\big[\mathrm{Tr}(\iota_I R_+ \cdot \iota_J R_+) - \mathrm{Tr}(\iota_I F \cdot \iota_J F) \big] = 0 \,. \quad (2.4.21)$$

We proceed now by inserting our ansatz (2.4.15) into the equations of motion. Since the three form flux H and the gauge fields are assumed to have no legs in the external four dimensions, the form of their EoM's stays exactly the same. One only has to keep in mind that one deals with six-dimensional fields and replace the ten-dimensional Hodge star by its six-dimensional counterpart. For the same reason the dilaton EoM also keeps its form, once the expansion of the ten-dimensional curvature scalar $\mathcal{R}_{X_{10}}$ and the curvature square $\mathrm{Tr}(R_+^{(10)})^2$

$$\mathcal{R}_{X_{10}} = e^{-2A_h} R_{X_4} + \mathcal{R} - 20\,|\mathrm{d}A_h|^2 - 8\nabla^2 A_h \,, \tag{2.4.22}$$

$$\mathrm{Tr}(R_+^{(10)})^2 = \mathrm{Tr}\, R_+^2 + 8\,e^{-2A_h}(\nabla_i \nabla_j e^{A_h})(\nabla^i \nabla^j e^{A_h}) + 4\,|\mathrm{d}A_h \lrcorner H|^2 \tag{2.4.23}$$
$$+ \frac{1}{12}\left| e^{-2A_h} R_{X_4} - 12|\mathrm{d}A_h|^2 \right|^2$$

is inserted. Demanding that the dilaton EoM is satisfied leaves us with the modified Einstein equation (2.4.21). Surprisingly, from this one can obtain rather restricting conditions on the heterotic warp factor A_h and the four-dimensional curvature scalar R_{X_4}, which is directly proportional to the cosmological constant for the maximally symmetric spaces we are considering. In particular, by choosing $I, J = \mu, \nu$ and contracting with the metric on X_4, one gets the equation

$$\nabla^i(e^{-2\Phi}\nabla_i e^{4A_h}) = e^{2A_h - 2\Phi} R_{X_4} + \alpha' e^{2A_h - 2\Phi} \Big\{ \frac{1}{24} e^{-2A_h}\Big| R_{X_4} - |\mathrm{d}e^{A_h}|^2 \Big|^2$$
$$+ 2(\nabla_i \nabla_j e^{A_h})(\nabla^i \nabla^j e^{A_h}) + |\mathrm{d}(e^{A_h}) \lrcorner H|^2 \Big\} \,. \tag{2.4.24}$$

Solving this equation perturbatively, we find that at the lowest order in α' we have

$$\nabla^i(e^{-2\Phi}\nabla_i e^{4A_h}) \simeq e^{2A_h - 2\Phi} R_{X_4} \,, \tag{2.4.25}$$

where by $\simeq$ we mean equivalence at zeroth-order in α'. By integrating this equation over the internal space, we get at the lowest order in α' the condition

$$R_{X_4} \int_{M_h} e^{2A_h - 2\Phi}\, \mathrm{vol}_{M_h} \simeq 0 \quad \Rightarrow \quad R_{X_4} \simeq 0 \,. \tag{2.4.26}$$

Note that multiplying (2.4.25) by $e^{(p-4)A_h}$ for any $p \neq 4$, integrating over M_h and using (2.4.26) one finds

$$\int_{M_h} \mathrm{vol}_{M_h}\, e^{p A_h - 2\Phi}\,(\mathrm{d}A_h)^2 \simeq 0 \,, \tag{2.4.27}$$

which implies that the string-frame warp factor must be constant at lowest order in α'.

Hence, at leading order in α', the modified external Einstein equations (2.4.21) implies that

R_{X_4} vanishes and e^A is constant. Plugging this back into (2.4.24) and expanding it in powers of α', one can check that the first corrections to the above result arise at order α'^3, and can thus be ignored at our level of accuracy.

We therefore conclude that, at order $\mathcal{O}(\alpha')$, the external Einstein equation (2.4.24) requires the four-dimensional space to be flat and the warping to be constant. Note that this result is valid for any purely bosonic compactification, whether it is supersymmetric or not.

We will now show that one arrives at the same results if one starts directly with the potential (2.4.16). To this end, we consider first the two equations of motion for the scalar fields Φ and A_h

$$\delta_\Phi: \quad \mathcal{R}_{X_{10}} - \frac{1}{2}|H|^2 + 4|\mathrm{d}\Phi|^2 - 4\nabla^2\Phi - 16\,\mathrm{d}A \lrcorner \mathrm{d}\Phi - \frac{\alpha'}{4}\big(\mathrm{Tr}(R_+^{(10)})^2 - \mathrm{Tr}\,F^2\big) = 0 \,, \tag{2.4.28}$$

$$\delta_{A_h}: \quad e^{-2A_h}\mathcal{R}_{X_4} - 2\mathcal{R}_{X_{10}} + 8|\mathrm{d}\Phi|^2 - 8\nabla^2\Phi + |H|^2 - 4\nabla^2 A_h - 16|\mathrm{d}A_h|^2 \tag{2.4.29}$$
$$- 24\,\mathrm{d}A_h \lrcorner \mathrm{d}\Phi - \alpha'\big(\mathrm{Tr}(R_+^{(10)})^2 - \mathrm{Tr}\,F^2\big) = \mathcal{O}(\alpha') \,.$$

Here, we used (2.4.22) and (2.4.23) to write the results more compactly. Furthermore, $\mathcal{O}(\alpha')$ on the right hand side of the warp factor EoM denotes all terms coming from the last line of (2.4.16). Inserting the dilaton EoM into the warp factor EoM gives us a condition similar to (2.4.24)

$$\nabla^i(e^{-2\Phi}\nabla_i e^{4A_h}) = e^{2A_h - 2\Phi} R_{X_4} + \mathcal{O}(\alpha') \,, \tag{2.4.30}$$

which yields again the result that $\mathrm{d}A_h$ and R_{X_4} are at least of order $(\alpha')^3$ and can therefore consistently be set to zero in our analysis. Applying this to the dilaton EoM (2.4.28) returns exactly the dilaton equation of motion derived directly from ten dimensions.

The result that the warp factor can be considered constant also simplifies the re-derivation of the other equations of motion. In particular, one can consistently neglect the last line of (2.4.16) when one varies with respect to the other fields. Also, the derivatives of A_h arising from partial integration can be set to zero. It is then clear by comparing the ten-dimensional action (2.2.1) and the potential (2.4.16) that the equations of motion for the two-form field B and the gauge potential $\mathcal{A}$ are identical irrespective whether they are derived from ten dimensions directly or via the potential (2.4.16).

The last step in showing the equivalence of the two approaches is then to consider variations of (2.4.16) with respect to the internal metric. These give

$$R_{ij} + 2\nabla_i\nabla_j\Phi - \frac{1}{2}\iota_i H \lrcorner \iota_j H + \frac{\alpha'}{4}\Big(\mathrm{Tr}\,\iota_i R_+ \lrcorner \iota_j R_+ - \mathrm{Tr}\,\iota_i F \lrcorner \iota_j F\Big) = \mathcal{O}(\alpha'^3) \,. \tag{2.4.31}$$

Exactly the same equation is obtained by reducing the modified Einstein equation (2.4.21) to the internal space (i.e. $I, J = i, j$). We conclude that also for compactifications of heterotic supergravity it is sufficient to consider an effective potential instead of the full ten-dimensional action.

Let us finish this section with some additional remarks. First, one should note that for both, the heterotic as well as the M-theory case, the potentials (2.4.9) and (2.4.16) must vanish on-shell. Both must be extremized under a general variation of the corresponding warp factor. In particular, they must be extremized by a constant shift of the warp factor, which demands that the potentials are zero on-shell.

Also, one might wonder why one finds a constant warp factor in the heterotic case, while this is not possible in eleven dimensions. This seeming discrepancy lies in the fact that we used the downstairs picture when discussing the EoM's of Horava-Witten theory. As mentioned, the EoM's should be accompanied by boundary conditions in the downstairs picture, which we did not derive explicitly. Presumably, these conditions will give extra requirements on the warp factor. However, one should be careful here, since not the warp factor A of (2.4.1) will be constant, but A', which appeared on the boundary. This can also be seen by fixing the overall normalization of the warpings by connecting them to the four-dimensional Planck mass[8] M_P

$$M_\text{P}^2 = \frac{1}{\kappa_{10}^2} \int_{M_h} \text{vol}_{M_h} e^{2A_h - 2\Phi} = \frac{1}{\kappa^2} \int_M \text{dvol}_7 \, e^{2A} \,. \qquad (2.4.32)$$

By reducing M-theory to heterotic SUGRA the integral over seven dimensions becomes

$$\frac{1}{\kappa^2} \int_M \text{dvol}_7 \, e^{2A} = \frac{2\pi\rho}{\kappa^2} \int_{M_h} \text{vol}_{M_h} e^{2A' - 2\Phi} \,, \qquad (2.4.33)$$

where we have set $\sigma = 2/3\,\Phi$ as it should be for a proper dimensional reduction, and thus we obtain $A_h = A' = A + \Phi/3$. This shows that indeed the field A' should be constant in the context of Horava-Witten theory.

Furthermore, as the warp factor is constant for heterotic SUGRA one can use (2.4.32) to connect it to the string coupling g_s

$$e^{2A_h} = \frac{g_s^2 l_s^8 M_\text{P}^2}{4\pi \text{Vol}(M)} \,, \qquad (2.4.34)$$

where $l_s = 2\pi\sqrt{\alpha'}$ is the string length scale and $\text{Vol}(M_h) = \int \text{vol}_{M_h}$ is the volume of the internal

---

[8]This is obtained by requiring that the dimensional reduction to four dimensions gives the canonical four-dimensional Einstein term $(M_\text{P}^2/2) \int \sqrt{-g_{X_4}} \, \mathcal{R}_{X_4}$.

space. Moreover, the string coupling g_s is defined by

$$\frac{1}{g_s^2} = \frac{\int_M e^{-2\Phi}\,\mathrm{vol}_M}{\int_M \mathrm{vol}_M} \, . \qquad (2.4.35)$$

This is all that we will need to discuss the supersymmetry properties of heterotic supergravity and heterotic M-theory in the following chapters. Most important for our discussion is the fact that we are allowed to use the derived potentials instead of the full action. However, in order to see how one can obtain information out of these potentials, we still need to introduce the concept of G-structures and its applications to flux compactifications. We will give a short introduction to these topics in the next chapter.

Chapter 3

G-Structures

The importance of G-structures for modern flux compactification cannot be underestimated, as was already demonstrated in the introduction. Here, we will focus on the technical aspects of G-structures and explain in more detail why they are so well suited to discuss supersymmetric string compactifications with or without flux. To this end we give a definition of G-structures in terms of the structure group of the tangent frame bundle of the concerned manifold and discuss the meaning of tensors (and spinors) invariant under the group G. As it turns out, the SUSY generator ϵ has to be an invariant spinor for supersymmetric compactifications. This implies that the supersymmetry variation of the gravitino can be used to determine the G-structure of the compactification manifold. Furthermore, on the internal manifold non-vanishing flux will lead naturally to a torsionful connection, which can be classified by using the appropriate G-structure. These two facts explain the significance of G-structures for flux compactifications. In chapter 4 and 5 we will make extensive use of G-structures. In particular we will need SU(3) structures in both six and seven dimensions, as well as G_2 structures in seven dimensions. We will therefore discuss these at the end of this chapter.

3.1 General remarks on G-structures

In this section we discuss common properties of G-structures and their relation to physics. We specify to SU(3) structures in six and seven dimensions and to G_2 structures in the next sections

3.1.1 An intuitive definition of G-structures

A quite intuitive definition and description of G-structures can be given in terms of the tangent frame bundle of the concerned manifold M.[1] Let us therefore remind the reader of the concept of a

---
[1] In our presentation we will follow closely [204], more material can be found, e.g. in [153, 154, 156]

bundle first. A generic bundle E consists of a base B and a fibre F such that locally it looks like the direct product of B and F. However, globally it does not need to have a direct product structure. This is taken into account by transition functions that determine how the fiber transforms if one goes from one patch on the base to another patch. Only when all transition functions equal the identity, one obtains globally the direct product manifold $B \times F$. Moreover, the bundle comes with a smooth projection π to the base. Assigning an element of the fiber to every point of the base, defines a section s of the bundle. The map s should of course satisfy the relation $\pi(s(p)) = p$ for every point $p \in B$.

A very simple example for a trivial bundle is the cylinder. There, the base space is taken to be the circle S^1 and the fiber is a finite line. A section of this bundle is given by a closed line that winds exactly once around the cylinder. A well known non-trivial generalization of this is the moebius strip. Here, the base and the fiber are the same as for the cylinder, but one has two patches whose transition function is an inversion of the line, resulting in the non-orientedness of the moebius strip. But still, a section is given by a closed line that winds once around the strip like in the case of the cylinder.

Another prominent class of examples for bundles, which is more important for our discussion, is provided by vector bundles. Here, the fiber is a vector space and a section of the bundle is given by a vector field. In particular, to every manifold M one can associate its tangent vector bundle (or tangent bundle, for short) TM. The fiber of this bundle at a point $p \in M$ is given by the vector space T_pM, that contains all vectors tangent to the point p. A basis for this space is given by the partial derivatives in every direction ∂_A. Clearly, for a d-dimensional manifold T_pM is a d-dimensional vector space and the transition functions of the fiber are given by elements of the group $\mathrm{GL}(d, \mathbb{R})$. A section assigns to every point p one vector of the space T_pM and gives hence a vector field over M.

The tangent frame bundle associated to TM can then be defined as the bundle whose fiber for a given point $p \in M$ is the set of ordered bases of the vector space T_pM. Locally, as we explained above, the bundle looks like the direct product $(p, e_{\underline{A}})$ with $e_{\underline{A}} = e^A{}_{\underline{A}} \partial_A$ a set of d independent vectors forming a base of T_pM. Note that the matrix $e^A{}_{\underline{A}}$ is only restricted by the condition that the $e_{\underline{A}}$ form a base of T_pM. Hence, $e_{\underline{A}}$ should really be understood as the set of all (ordered) bases.

What can one say about the transition functions of the tangent frame bundle between two different overlapping patches U_α and U_β with coordinates x^A and x'^A, respectively? In the overlap region one can represent $e_{\underline{A}}$ in terms of both coordinates

$$e_{\underline{A}} = (e_\alpha)^A{}_{\underline{A}} \partial_A = (e_\beta)^A{}_{\underline{A}} \partial'_A = (e_\beta)^B{}_{\underline{A}} \frac{\partial x^A}{\partial x'^B} \partial_A = (e_\beta)^B{}_{\underline{A}} (t_{\beta\alpha})_B{}^A \partial_A \,. \tag{3.1.1}$$

Moreover, the transition functions $t_{\beta\alpha}$ must satisfy the consistency condition

$$t_{\alpha\beta} t_{\beta\alpha} = 1 \,, \tag{3.1.2}$$

and the transitivity relation on an overlap region of three patches U_α, U_β, U_γ

$$t_{\alpha\beta} t_{\beta\gamma} = t_{\alpha\gamma} \,. \tag{3.1.3}$$

All these requirements provide a group structure for the transition functions $t_{\alpha\beta}$. This group is called the structure group of M. In general it will be the group of general linear transformations in d dimensions $\mathrm{GL}(d,\mathbb{R})$, and the same group as for the transition functions of the tangent bundle TM.

After this preliminary work, it is quite easy to define G-structure. A manifold is said to be a G-structure manifold if its structure group can be reduced to the group G. Differently put: a manifold will provide a G-structure when the transition functions of the tangent frame bundle belong to the group G. But since every vector l from the tangent bundle can be decomposed as $l = l^A e_A$ this means that the vectors of the tangent bundle transform under the group G, too. This in turn implies that also one-forms and generic tensors transform under G when one considers a G-structure manifold.

The next question is then of course what the structure group of a given manifold is. Again, one can find the answer by considering tensors and not the full tangent frame bundle. As a matter of fact a tensor (or a spinor) that is globally well defined (and non-degenerate) on M will reduce the structure group. The simplest way to imagine this is to assume that one has already found as structure group the d dimensional rotations. If one finds now in addition a nowhere vanishing vector l the group will reduce to the $(d-1)$-dimensional rotations, since one can find frames such that l points always in the same direction.

Another well known case with reduced structure group are Riemannian manifolds, e.g. manifolds that admit a metric g. The structure group is then reduced to the orthogonal transformations, as the metric fixes the length of vectors in all patches. Also important is the case where one can define an almost complex structure on the manifold M. This is a map $J : TM \to TM$, that squares to minus one: $J^2 = -1$. This means that J has eigenvalues of $+i$ and $-i$ and hence the structure group will reduce to $\mathrm{GL}(d/2, \mathbb{C})$. If one has both, a metric g and an almost complex structure J, that satisfy $JgJ = g$ one speaks of a hermitian metric. The structure group is then given by the unitary group $\mathrm{U}(d/2)$. In this case one can also define a two-form J as

$$J_{ij} = g_{ik} J^k{}_j \,. \tag{3.1.4}$$

This two-form is called pre-symplectic structure[2] and it would reduce the structure group to the symplectic group $\mathrm{Sp}(d,\mathbb{R})$ if considered solely.

3.1.2 G-structures and torsion

In the last section we used invariant tensors to determine the structure group of M. However, it might be quite difficult to be sure if one has really found all invariant tensors in some cases. Therefore, it is more systematically to start with a given G-structure and ask then which invariant tensors can be built. Following [155, 159, 205–208] we will describe how to do this and how one can further classify G-structure manifolds by torsion.

As we have seen the most generic structure group of a d-dimensional manifold is $\mathrm{GL}(d,\mathbb{R})$. This is true as long as one does not want to have spinors included. However, in supergravity theories it is mandatory to include spinors, and hence the manifolds considered should be suitable for this. On the other hand, we will only deal with metric spaces such that the maximal structure group will be reduced to $\mathrm{O}(d,\mathbb{R})$. In order to define spinor bundles on M it must be possible to define the double cover $\mathrm{Spin}(d)$ of $\mathrm{O}(d,\mathbb{R})$ in a globally consistent way. We conclude that the manifolds that can be considered in supergravity compactifications should at least have structure group $\mathrm{Spin}(d)$.

It is then not difficult to find the invariant tensors for a given G-structure. One simply has to decompose the tensor representations of $\mathrm{Spin}(d)$ with respect to G, to see whether there are invariants. For example, considering a seven dimensional manifold with $\mathrm{SU}(2)$ structure we find one real and one complex vector, respectively, which are invariant under $\mathrm{SU}(2)$, by decomposing the fundamental **7** of $\mathrm{Spin}(7)$

$$\mathbf{7} \to \mathbf{1} + (\mathbf{1} + \mathbf{2}) + (\bar{\mathbf{1}} + \bar{\mathbf{2}}) \,. \tag{3.1.5}$$

We will discuss more such decompositions later on when we consider the cases relevant for our work.

Given now that for a structure group G there is an invariant tensor Υ, one can further classify the structure by considering the covariant derivatives of Υ. One starts by defining a connection $\nabla^{(T)}$ that satisfies $\nabla^{(T)}\Upsilon = 0$. As indicated by the T this connection will in general have intrinsic torsion τ that is measured by the difference of $\nabla^{(T)}$ and the Levi-Civita connection ∇ associated with the metric on M

$$\tau = \nabla - \nabla^{(T)} \,. \tag{3.1.6}$$

Being the difference between two connections the torsion is a tensor that takes values in $\Lambda^1 \otimes \Lambda^2$.[3]

---
[2]Giving the same name to the pre-symplectic structure J_{ij} and the almost complex structure $J^i{}_j$ is commonly done in the literature.
[3]Λ^p is the space of p-forms.

A more detailed classification of the G-structure manifolds can then be obtained by decomposing τ with respect to G and by analyzing which G-modules are present within τ, and which are not, respectively.

In order to do so, one notices that in d dimensions Λ^2 is isomorphic to the Lie algebra of $SO(d)$, which in turn can be split into the Lie algebra of G and its orthogonal complement[4]

$$\Lambda^2 \cong \mathfrak{so}(d) = \mathfrak{g} \oplus \mathfrak{g}^\perp \ . \tag{3.1.7}$$

However, since Υ is invariant under G, it is also invariant under $\mathfrak{g}$. The action of τ on Υ is therefore given by $\Lambda^1 \otimes \mathfrak{g}^\perp$. The G-modules appearing in $\Lambda^1 \otimes \mathfrak{g}^\perp$ are called torsion classes, and provide the more detailed classification of G-structure manifolds. If the torsion τ is identically zero the manifold is said to have G-holonomy instead of only G-structure. We will use this classification later on when discussing $SU(3)$ and G_2 structures.

3.1.3 G-structures and supersymmetry

In the previous sections we have given an intuitive definition of G-structures and a classification in terms of invariant tensors and derivatives of these tensors. One may ask then: what is the connection of these concepts to physics? The answer to this question is provided by not only considering invariant tensors, but also spinors. In fact, as we mentioned before, supergravity theories demand at least one globally defined non-vanishing spinor, i.e. a spinor that is invariant under the structure group G. But such a spinor ε should also satisfy

$$\nabla_A^{(T)} \varepsilon = \left(\nabla_A - \frac{1}{4}\tau_A{}^{BC}\tilde{\Gamma}_{BC}\right)\varepsilon = 0 \ . \tag{3.1.8}$$

It is obvious that the gravitino variation of eleven dimensional SUGRA (2.1.4) and of heterotic SUGRA (2.2.6a) are exactly of this form. More general, the gravitino variations of all ten and eleven dimensional supergravity theories are of the form (3.1.8). Comparing (3.1.8) to (2.1.4) and (2.2.6a) it can be understood that the torsion will depend on the flux which appears in the gravitino variation.

The most prominent case of this interdependence of torsion and flux is the case of Calabi-Yau compactification, firstly described in the seminal papers [27, 130]. There, it was shown that in the absence of flux the internal manifold of a heterotic string compactification has to have $SU(3)$ holonomy, i.e. it has to be a Calabi-Yau manifold, while for non-vanishing flux only an $SU(3)$ structure is possible. Later on this kind of analysis has been extended to the other sectors of string

---

[4]Since the analysis is on the level of Lie algebras here, which are connected to the identity, one is allowed to consider the Lie algebras of $SO(d)$ instead of $Spin(d)$.

theory and also to eleven dimensional supergravity.

Although equation (3.1.8) contains all information that is needed to classify the compactification manifold by torsion classes, the information is quite entangled. A better representation can be obtained by employing that the invariant p-forms of a given G-structure can be constructed out of invariant spinors of the same structure by building Clifford-algebra scalars

$$\Upsilon_p = \frac{1}{p!} \left(\varepsilon^\dagger \tilde{\Gamma}_{A_1 \dots A_p} \varepsilon \right) \mathrm{d}x^{A_1 \dots A_p}, \qquad \tilde{\Upsilon}_p = \frac{1}{p!} \left(\varepsilon^T \tilde{\Gamma}_{A_1 \dots A_p} \varepsilon \right) \mathrm{d}x^{A_1 \dots A_p}. \qquad (3.1.9)$$

Clearly, since $\nabla_A^{(T)} \varepsilon = 0$, also $\nabla_A^{(T)} \Upsilon_p = \nabla_A^{(T)} \tilde{\Upsilon}_p = 0$ as it should be for an invariant form. This means that one can classify the torsion of the G-structure by considering the exterior derivatives of the invariant forms Υ_p and $\tilde{\Upsilon}_p$. The G-modules which appear in the decomposition of these derivatives are in one to one correspondence to the torsion classes of the manifold. Since the supersymmetry variation of the gravitino has to vanish in order to provide supersymmetric vacua, and since this variation has exactly the form of (3.1.8), one can use the exterior derivatives of the invariant p-forms to establish criteria that a compactification manifold has to fulfill in order to give supersymmetric vacua. This is the reasons why G-structures are so useful in the discussion of flux compactification. In chapter 4 and chapter 5 we will use these results in our analysis. But before we do so, we need to give a more thorough account of the used G-structures, that is of SU(3) structures in six and seven dimensions and of G_2 structures in seven dimensions.

3.2 SU(3) structures in six dimensions

We start our discussion with SU(3) structures in six dimensions (see e.g. [1, 134, 150, 159, 204, 206, 207]). This case will be relevant for heterotic supergravity, since compactification to four dimensions lead to a six-dimensional internal manifold M with structure group SO(6). This group is isomorphic to SU(4) and its **4** is the irreducible chiral spinor representation of SO(6). In order to have $\mathcal{N} = 1$ supersymmetry in four dimensions, there should be only one invariant spinor on M. Decomposing the **4** with respect to various subgroups of SU(4), one finds that SU(3) gives one invariant spinor. Hence, in order to end up with $\mathcal{N} = 1$ SUSY in four dimension, one has to compactify on SU(3) structure manifolds. Scanning through the smallest irreducible representations of SU(4) one finds the following decompositions with respect to SU(3)

$$\begin{aligned}
\text{spinor}: \quad & \mathbf{4} \to \mathbf{1} + \mathbf{3} & (3.2.1)\\
\Lambda^1: \quad & \mathbf{6} \to \mathbf{3} + \bar{\mathbf{3}} & \\
\Lambda^3/2: \quad & \mathbf{10} \to \mathbf{1} + \mathbf{3} + \mathbf{6} & \\
\Lambda^2: \quad & \mathbf{15} \to \mathbf{1} + \mathbf{3} + \bar{\mathbf{3}} + \mathbf{8}\,.
\end{aligned}$$

Note that Λ^3 would be the **20** of SO(6). By $\Lambda^3/2$ we denote 3-forms that can locally be promoted to holomorphic forms. Their representation is therefore the **10**. We see that we have besides one invariant chiral spinor η_+ one invariant two-form J and one invariant three-form Ω. In a moment we will explain how one can interpret these forms as an almost complex structure and an associated $(3,0)$-form on M.

As discussed in the last section the connection between η_+, J, and Ω is established by

$$J_{ij} = \frac{i}{\|\eta_+\|^2}\,\eta_+^\dagger \gamma_{ij} \eta_+\,, \qquad \Omega_{ijk} = \frac{1}{\|\eta_+\|^2}\,\eta_+^T \gamma_{ijk} \eta_+\,, \qquad (3.2.2)$$

where $\|\eta_+\|^2 = \eta_+^\dagger \eta_+$. Making use of the Fierz identities

$$\eta_+ \eta_+^\dagger = \frac{1}{8}\sum_{l=0}^{6} \frac{1}{l!}\,\eta_+^\dagger \gamma_{i_l\ldots i_1} \eta_+\, \gamma^{i_1\ldots i_l}\,, \qquad \eta_+ \eta_+^T = \frac{1}{8}\sum_{l=0}^{6} \frac{1}{l!}\,\eta_+^T \gamma_{i_l\ldots i_1} \eta_+\, \gamma^{i_1\ldots i_l}\,, \qquad (3.2.3)$$

one can show that J and Ω define indeed an almost complex structure[5]

$$J_i{}^j J_j{}^k = -\delta_i{}^k\,, \qquad J_i{}^j \Omega_{jkl} = -i\,\Omega_{ikl}\,. \qquad (3.2.4)$$

Here it is worth to notice that in the expansion of $\eta_+ \eta_+^\dagger$ only even numbers of gamma matrices contribute, while for $\eta_+ \eta_+^T$ the only contributions comes from $l = 3$. Defining the local projection operator on $(1,0)$-forms

$$P_i{}^j = \frac{1}{2}(\delta_i{}^j + i J_i{}^j)\,, \qquad (3.2.5)$$

one finds that locally J is a $(1,1)$-form and Ω a $(3,0)$-form respectively. As such they satisfy the relations

$$J \wedge \Omega = 0\,, \qquad \frac{1}{3!} J \wedge J \wedge J = -\frac{i}{8}\,\Omega \wedge \overline{\Omega}\,. \qquad (3.2.6)$$

These relations will hold not only locally, but globally, when the almost complex structure is

---

[5]To show this one has to use relations like $\eta_+^\dagger \eta_+ \eta_+^\dagger \eta_+ = \|\eta_+\|^4$ and the Fierz identities, and compare coefficients of gamma matrices.

integrable. This is the case when the exterior derivative of Ω is proportional to itself, i.e. when $\mathrm{d}(e^{Y_1}\Omega) = 0$ for some scalar function Y_1. Then, the SU(3) structure manifold will be a complex manifold. In contrast, when $\mathrm{d}(e^{Y_2}J) = 0$ one speaks of a Kähler manifold. Finally a manifold that satisfies both of these conditions is a Calabi-Yau manifold.

Making use of (A.1.12) one can calculate the action of the Hodge star on J and Ω

$$*_6 J = \frac{1}{2} J \wedge J, \quad *_6 \Omega = -i\,\Omega\,, \qquad (3.2.7)$$

which shows that Ω is imaginary anti-self-dual. Furthermore, one can also construct the projection operator (3.2.5) from the components of Ω

$$\Omega_{ikl}\,\overline{\Omega}^{jkl} = 16\,P_i{}^j = 8\,(\delta_i{}^j + i\,J_i{}^j)\,. \qquad (3.2.8)$$

The torsion classes can be expressed in terms of the exterior derivatives of J and Ω. The torsion $\tau \in \Lambda^1 \otimes \mathfrak{g}^\perp = \mathbf{6} \otimes (\mathbf{15} - \mathfrak{g})$ decomposes with respect to SU(3) as

$$\begin{aligned}
(\mathbf{3}+\bar{\mathbf{3}}) \otimes (\mathbf{1}+\mathbf{3}+\bar{\mathbf{3}}) &= (\mathbf{1}+\mathbf{1}) + (\mathbf{8}+\mathbf{8}) + (\mathbf{6}+\bar{\mathbf{6}}) + (\mathbf{3}+\bar{\mathbf{3}}) + (\mathbf{3}+\bar{\mathbf{3}}) \\
&= W_1 + W_2 + W_3 + W_4 + (W_5 + \overline{W}_5)\,.
\end{aligned} \qquad (3.2.9)$$

In terms of the torsion classes W_i, $\mathrm{d}J$ and $\mathrm{d}\Omega$ read

$$\mathrm{d}J = -\frac{3}{2}\mathrm{Im}(\overline{W}_1\Omega) + W_4 \wedge J + W_3\,, \qquad (3.2.10)$$

$$\mathrm{d}\Omega = W_1\,J \wedge J + W_2 \wedge J + \overline{W}_5 \wedge \Omega\,, \qquad (3.2.11)$$

where W_1 is a complex scalar, W_2 is a primitive (1,1)-form, and W_3 is real $(2,1) + (1,2)$ and primitive. W_4 is a real one-form and W_5 is a (1,0)-form.[6] These are the basic properties of SU(3) structures in six dimensions that are needed for the work presented in chapter 4. We will therefore turn to seven dimensions now.

3.3 G_2 and SU(3) structure in seven dimensions

In this section we will review some points concerning G_2 and SU(3) structures on seven dimensional manifolds which will become important in chapter 5 (see e.g. [206–208, 210–217]). For a seven dimensional manifold the structure group to start with is SO(7) (or its double cover Spin(7)). Scanning the decomposition of the spinor representation $\mathbf{8}$ with respect to subgroups of SO(7) for

---

[6] A notion of primitivity, which is sufficient for our purposes, is that a form is primitive if its contractions with J and Ω vanishes. For more information on the topic see e.g. [10, 209].

one invariant spinor, one is lead to the exceptional group G_2. The one-, two-, and three-forms decompose according to

$$\begin{aligned} \text{spinor}: & & \mathbf{8} &\to \mathbf{1} + \mathbf{7}, \\ \Lambda^1: & & \mathbf{7} &\to \mathbf{7}, \\ \Lambda^2: & & \mathbf{21} &\to \mathbf{7} + \mathbf{14}, \\ \Lambda^3: & & \mathbf{35} &\to \mathbf{1} + \mathbf{7} + \mathbf{27}. \end{aligned} \qquad (3.3.1)$$

Hence, a G_2 structure manifold is completely determined by an invariant three-form ϕ, or equivalently by a globally well defined SO(7) Majorana spinor η. Normalizing this spinor such that $\|\eta\|^2 = 1$ one can relate these quantities by

$$\phi = \frac{i}{3!} \eta^\dagger \gamma_{mnp} \eta \, dx^{mnp}, \qquad *_7 \phi = \psi = -\frac{1}{4!} \eta^\dagger \gamma_{mnpq} \eta \, dx^{mnpq}. \qquad (3.3.2)$$

For a manifold of G_2 holonomy, $d\phi = d\psi = 0$ would hold. The departure from holonomy can be measured by the G_2 torsion classes

$$d\phi = \tau_0 \psi + 3\tau_1 \wedge \phi + *_7 \tau_3, \qquad d\psi = 4\tau_1 \wedge \psi + \tau_2 \wedge \phi. \qquad (3.3.3)$$

These torsion classes are elements of $\Lambda^1 \otimes \mathfrak{g}^\perp = \mathbf{7} \otimes (\mathbf{21} - \mathfrak{g})$

$$\begin{aligned} \mathbf{7} \otimes \mathbf{7} &= \mathbf{1} + \mathbf{7} + \mathbf{14} + \mathbf{27} \\ &= \tau_1 + \tau_2 + \tau_3 + \tau_4. \end{aligned} \qquad (3.3.4)$$

By inverting (3.3.3) it is possible to express them in terms of the invariant forms ϕ and ψ

$$\begin{aligned} \tau_0 &= \frac{1}{7} d\phi \lrcorner \psi, & \tau_1 &= -\frac{1}{12} d\phi \lrcorner \phi = \frac{1}{12} d\psi \lrcorner \psi, \\ \tau_2 &= \frac{1}{2} (d\psi \lrcorner \phi - *_7 d\psi) - 2\tau_1 \lrcorner \phi & \tau_3 &= *_7 d\phi - \tau_0 \phi + 3\tau_1 \lrcorner \psi. \\ &= -*_7 d\psi + 4\tau_1 \lrcorner \phi, \end{aligned} \qquad (3.3.5)$$

A set of extremely useful identities can be obtained by choosing locally an explicit representation of ϕ, e.g. [210]

$$\begin{aligned} \phi &= dx^{\underline{123}} + dx^{\underline{145}} + dx^{\underline{167}} + dx^{\underline{246}} - dx^{\underline{257}} - dx^{\underline{347}} - dx^{\underline{356}} = \frac{1}{3!} \phi_{mnp} dx^{mnp}, \\ \psi &= dx^{\underline{4567}} + dx^{\underline{2367}} + dx^{\underline{2345}} + dx^{\underline{1357}} - dx^{\underline{1346}} - dx^{\underline{1256}} - dx^{\underline{1247}} = \frac{1}{4!} \psi_{mnpq} dx^{mnpq}. \end{aligned} \qquad (3.3.6)$$

Using this representation one can show by explicit calculations that ϕ and ψ satisfy the relations

$$\phi_{mnp}\, \phi^{mqr} = \psi_{np}{}^{qr} + 2\, \delta^q_{[n}\, \delta^r_{p]}\,, \qquad (3.3.7)$$

$$\phi_{mnp}\, \psi^{mqrs} = 6\, \delta^{[q}_{[n}\, \phi_{p]}{}^{rs]}\,,$$

$$\psi_{mnpq}\, \psi^{mnrs} = 2\psi_{pq}{}^{rs} + 8\, \delta^r_{[p}\, \delta^s_{q]}\,.$$

In order to obtain an SU(3) structure from a G_2 structure, one needs a globally well defined invariant one-form v as can be seen by further decomposing (3.3.1) with respect to SU(3)

$$\begin{aligned}
\text{spinor}: &\quad \mathbf{8} \to \mathbf{1} + \bar{\mathbf{1}} + \mathbf{3} + \bar{\mathbf{3}}\,,\\
\Lambda^1: &\quad \mathbf{7} \to \mathbf{1} + \mathbf{3} + \bar{\mathbf{3}}\,,\\
\Lambda^2: &\quad \mathbf{21} \to \mathbf{1} + 2\times\mathbf{3} + 2\times\bar{\mathbf{3}} + \mathbf{8}\,,\\
\Lambda^3: &\quad \mathbf{35} \to \mathbf{1} + \bar{\mathbf{1}} + \mathbf{1} + 2\times\mathbf{3} + 2\times\bar{\mathbf{3}} + \mathbf{6} + \bar{\mathbf{6}} + \mathbf{8}\,.
\end{aligned} \qquad (3.3.8)$$

Besides the one-form one finds two invariant spinors, one two-form, a complex three-form, and a real three-form. The first of the two spinors is the original G_2 spinor η, while the second spinor can be defined with the help of the one-form v

$$\eta_+ = \frac{1}{\sqrt{2}}\, e^{\frac{Z}{2}} (1 + v_m \gamma^m)\, \eta\,, \qquad \eta_+^* = \eta_- = \frac{1}{\sqrt{2}}\, e^{\frac{Z}{2}} (1 - v_m \gamma^m)\, \eta\,. \qquad (3.3.9)$$

Here Z is a real function and v_m denotes the components of v. Using these spinors one can construct several new forms on the SU(3) structure manifold

$$\Sigma_p = \frac{1}{p!}\, \eta_+^\dagger\, \gamma_{n_1\ldots n_p}\, \eta_+\, \mathrm{d}x^{n_1\ldots n_p}\,, \qquad \tilde{\Sigma}_p = \frac{1}{p!}\, \eta_+^T\, \gamma_{n_1\ldots n_p}\, \eta_+\, \mathrm{d}x^{n_1\ldots n_p}\,. \qquad (3.3.10)$$

Denoting the invariant two and three-forms with J, Ω, and $v \wedge J$ in analogy to the six dimensional SU(3) structure case the following relations can be established

$$\begin{aligned}
&\Sigma_0 = \|\eta_+\|^2 = \|\eta_-\|^2 = e^Z\,, &&\Sigma_7 = i\, e^Z\, \mathrm{dvol}_7\,, &&\tilde{\Sigma}_0 = \tilde{\Sigma}_1 = \tilde{\Sigma}_2 = 0\,,\\
&\Sigma_1 = e^Z\, v\,, &&\Sigma_6 = \frac{i}{3!}\, e^Z\, J \wedge J \wedge J\,, &&\tilde{\Sigma}_5 = \tilde{\Sigma}_6 = \tilde{\Sigma}_7 = 0\,,\\
&\Sigma_2 = -i\, e^Z\, J\,, &&\Sigma_5 = -\frac{1}{2}\, e^Z\, v \wedge J \wedge J\,, &&\tilde{\Sigma}_3 = -i\, e^Z\, \Omega\,,\\
&\Sigma_3 = -i\, e^Z\, v \wedge J\,, &&\Sigma_4 = -\frac{1}{2}\, e^Z\, J \wedge J\,, &&\tilde{\Sigma}_4 = i\, e^Z\, v \wedge \Omega\,.
\end{aligned} \qquad (3.3.11)$$

A detailed calculation shows that J and Ω satisfy relations analogous to the SU(3) structure relations in six dimensions

$$J \wedge \Omega = 0, \qquad \mathrm{dvol}_7 = v \wedge \mathrm{dvol}_6 = \frac{1}{3!} v \wedge J \wedge J \wedge J = -\frac{i}{8} v \wedge \Omega \wedge \bar{\Omega}. \qquad (3.3.12)$$

Furthermore, v is perpendicular to J and Ω

$$J \lrcorner v = 0, \qquad \Omega \lrcorner v = 0, \qquad (3.3.13)$$

and thus M looks locally like the direct product of a six dimensional SU(3) structure manifold and a line. It is then possible to use J and Ω in order to introduce an almost complex structure on this six dimensional subspace, since they satisfy

$$J_m{}^n J_n{}^p = -\delta_m^p + v_m v^p, \qquad \Omega_{mpq} \bar{\Omega}^{npq} = 8 \left(\delta_m^n + i J_m{}^n - v_m v^n \right), \qquad (3.3.14)$$
$$J_m{}^n \Omega_{npq} = -i \Omega_{mpq},$$

which means that Ω is $(3,0)$ and J $(1,1)$ with respect to this almost complex structure. Similarly to the six dimensional case, one can use the property (A.1.6) of the gamma matrices to calculate the Hodge duals of v, J, and Ω

$$*_7 v = \frac{1}{3!} J \wedge J \wedge J, \qquad *_7 J = \frac{1}{2} v \wedge J \wedge J, \qquad *_7 \Omega = -i v \wedge \Omega. \qquad (3.3.15)$$

Writing the G_2 spinor η in terms of η_+, one gets from (3.3.2) the connection between the G_2 and the SU(3) structure forms

$$\phi = v \wedge J + \mathrm{Re}\,\Omega, \qquad \psi = \frac{1}{2} J \wedge J + v \wedge \mathrm{Im}\,\Omega. \qquad (3.3.16)$$

The torsion is already quite complicated for this case, since it takes values in $\Lambda^1 \otimes \mathfrak{g}^\perp = (\mathbf{1} + \mathbf{3} + \bar{\mathbf{3}}) \otimes (\mathbf{1} + 2 \times \mathbf{3} + 2 \times \bar{\mathbf{3}})$. The departure from SU(3) holonomy is measured by 14 torsion classes

$$\begin{aligned}
dv &= R J + \bar{V}_1 \lrcorner \Omega + V_1 \lrcorner \bar{\Omega} + v \wedge W_0 + T_1, \\
dJ &= -\frac{3}{2} \mathrm{Im}(\bar{W}_1 \Omega) + W_4 \wedge J + W_3 + v \wedge \left(\frac{2}{3} \mathrm{Re} E\, J + \bar{V}_2 \lrcorner \Omega + V_2 \lrcorner \bar{\Omega} + T_2 \right), \\
d\Omega &= W_1 J \wedge J + W_2 \wedge J + \bar{W}_5 \wedge \Omega + v \wedge (E \Omega - 4 V_2 \wedge J + S).
\end{aligned}$$
(3.3.17)

Here, R is a real scalar, while W_1 and E are complex scalars. W_5, V_1, and V_2 are $(1,0)$-forms, while W_0 and W_4 are real one-forms. W_2, T_1, T_2 are primitive and $(1,1)$. W_3 and S are $(2,1)+(1,2)$ and primitive. All degrees of the forms are understood with respect to the almost complex structure defined by J and Ω. Note that while W_1 to W_5 are also present in the six dimensional case, the other torsion classes are special to seven dimensions and describe the embedding of the SU(3) structure manifold into M.

With the tools presented in this chapter, we are now ready to turn back to the physical problems we want to address. In the next chapter we will heavily rely on the results on SU(3) structures in six dimensions in order to deal with heterotic supergravity, while in chapter 5 seven dimensional G-structures will be needed for our discussion of heterotic M-theory.

Chapter 4

Heterotic Domain Wall Supersymmetry Breaking

After we have laid the theoretical foundations in the previous two chapters, we can now go on to the main topics of this thesis. In this chapter we will be concerned with the construction of non-supergravity vacua for heterotic supersymmetry. To this end, we will first rewrite the potential (2.4.16)

$$\begin{aligned}
V_h &= -\frac{1}{2\kappa_{10}^2}\int_{M_h} \text{dvol}_{M_h}\, e^{4A_h - 2\Phi}\left\{e^{-2A_h}R_{X_4} + \mathcal{R} - \frac{1}{2}H^2 - 4(\text{d}\Phi)^2 - 8\nabla^2 A_h\right. \\
&\quad - 20(\text{d}A_h)^2 + \frac{\alpha'}{4}(\text{Tr}\,R_+^2 - \text{Tr}\,F^2) \\
&\quad \left. + \alpha'\left[2e^{-2A_h}(\nabla^i\nabla^j e^{A_h})(\nabla_i\nabla_j e^{A_h}) + |\text{d}A_h \lrcorner H|^2 + 3|\frac{1}{12}e^{-2A_h}R_{X_4} - \text{d}A_h \lrcorner \text{d}A_h)^2|^2\right]\right\},
\end{aligned}$$

and the supersymmetry conditions (2.2.6)

$$\delta\Psi_I = \left(\nabla_I - \frac{1}{4}\slashed{H}_I\right)\epsilon = \nabla_I^- \epsilon = 0,$$

$$\delta\lambda = \left(\slashed{\partial}\Phi - \frac{1}{2}\slashed{H}\right)\epsilon = 0,$$

$$\delta\chi = \frac{1}{2}\slashed{F}\epsilon = 0,$$

in terms of the SU(3) structure forms J and Ω. By this we will see that the potential can be put into a BPS-like form, i.e. it becomes a sum of squares that vanish once SUSY is satisfied.

By a special pattern of controlled SUSY-breaking, called domain wall SUSY-breaking (DWSB), we will obtain severe constraints on the compactification manifold. By solving these constraints we will explicitly construct examples of DWSB-vacua. At the end of the chapter we will also discuss the effects that the inclusion of a gaugino condensate has on our vacua. We find that both

DWSB and gaugino condensation give rise to supersymmetry breaking and that one can encompass most of the non-supersymmetric vacua that are known up to now in the literature by these two SUSY-breaking patterns. The work presented in this chapter was firstly published in [1–3], and we will follow closely the presentation given there. Note that since we are concerned here only with ten-dimensional heterotic supergravity compactified on a six-dimensional manifold, we will denote the heterotic warp factor by A instead of A_h. Furthermore, the spinor ϵ is a Majorana-Weyl spinor of SO(1,9) and the Hodge star $*$ will be the six-dimensional Hodge star. We use this notation in order to keep the equations as simple as possible.

4.1 BPS-like potential and SUSY conditions

The scalar potential V written in the form (2.4.16) does not appear particularly useful to study supersymmetry. In order to improve the situation, we have to make manifest the underlying supersymmetric structure. As we are going to show, this is possible due to the use of the SU(3) structure of the internal space, its invariant (3,0)-form Ω, its Kähler (or fundamental) (1,1)-form J, and its related globally defined chiral spinor η_+, which were discussed in section 3.2. In supersymmetric compactifications [130, 134] η_+ can be seen as the internal component of the ten-dimensional SUSY-generator ϵ, that decomposes as

$$\epsilon = \zeta \otimes \eta_+ + \text{c.c.} . \qquad (4.1.1)$$

Here both ζ and η_+ are chiral in four and six dimensions, respectively: $\gamma_{(4)}\zeta = \zeta$ and $\gamma_{(6)}\eta_+ = \eta_+$. Note, that if X_4 is Minkowski space, then ζ is a constant chiral spinor. If X_4 is an AdS$_4$ space, then ζ is the Killing spinor defined by

$$\nabla_\mu \zeta = \frac{1}{2}\overline{w}_0 \hat{\gamma}_\mu \zeta^* , \qquad (4.1.2)$$

where w_0 has an arbitrary phase and is related to the AdS$_4$ radius R by $|w_0| = 1/R$. Let us stress that, at this stage, we do not require ϵ or η_+ to satisfy any particular requirements except being globally well-defined. In other words, Ω and J define a generic SU(3) structure.

The potential in terms of SU(3) structure

The key point for our analysis is that Ω and J also specify the six-dimensional metric g. Thus, in principle, one can express the scalar curvature $\mathcal{R}$ appearing in the potential (2.4.16) as a function of Ω and J. This problem has been addressed in [218], and in [197, 219] for the more general SU(3)×SU(3) structure case relevant in type II configurations. Here we use the general formula obtained in [197] (see equation (C.1) therein), from which one can derive the following identity[1]

$$\mathcal{R} = -\frac{1}{2}(\mathrm{d}J)^2 - \frac{1}{8}[\mathrm{d}(J \wedge J)]^2 - \frac{1}{2}|\mathrm{d}\Omega|^2 + \frac{1}{2}|J \wedge \mathrm{d}\Omega|^2 + \frac{1}{2}u^2 - \nabla^i u_i, \qquad (4.1.3)$$

with

$$u = u_i \mathrm{d}y^i = \frac{1}{4}(J \wedge J) \lrcorner \mathrm{d}(J \wedge J) - \frac{1}{2}\mathrm{Re}(\overline{\Omega} \lrcorner \mathrm{d}\Omega) \,. \qquad (4.1.4)$$

Thus, by using (4.1.3), together with the BI (2.2.5), one can rewrite (2.4.16) as[2]

$$V = V_0 + V_1, \qquad (4.1.5)$$

with

$$\begin{aligned}
V_0 &= \frac{1}{4\kappa_{10}^2} \int \mathrm{dvol}_M \, e^{4A-2\Phi} \big[e^{-4A+2\Phi} \mathrm{d}(e^{4A-2\Phi} J) - *H \big]^2 \\
&\quad + \frac{1}{4\kappa^2} \int \mathrm{dvol}_M \, e^{4A-2\Phi} \Big\{ \frac{1}{4} \big[e^{-2A+2\Phi} \mathrm{d}(e^{2A-2\Phi} J \wedge J) \big]^2 + 4(\mathrm{d}A)^2 \Big\} \\
&\quad + \frac{1}{4\kappa^2} \int \mathrm{dvol}_M \, e^{-2A+2\Phi} \Big[|\mathrm{d}(e^{3A-2\Phi}\Omega)|^2 - |J \wedge \mathrm{d}(e^{3A-2\Phi}\Omega)|^2 \Big] \\
&\quad - \frac{1}{4\kappa^2} \int \mathrm{dvol}_M \, e^{4A-2\Phi} \Big\{ 2\mathrm{d}A + \frac{1}{4} e^{-2A+2\Phi} (J \wedge J) \lrcorner \mathrm{d}(e^{2A-2\Phi} J \wedge J) \\
&\quad \qquad + \frac{1}{2} e^{-3A+2\Phi} \mathrm{Re}[\overline{\Omega} \lrcorner \mathrm{d}(e^{3A-2\Phi}\Omega)] \Big\}^2, \qquad (4.1.6a) \\
V_1 &= \frac{\alpha'}{8\kappa_{10}^2} \int e^{4A-2\Phi} \big[\mathrm{Tr}(F \wedge *F) + \mathrm{Tr}(F \wedge F) \wedge J \big] \\
&\quad - \frac{\alpha'}{8\kappa^2} \int e^{4A-2\Phi} \big[\mathrm{Tr}(R_+ \wedge *R_+) + \mathrm{Tr}(R_+ \wedge R_+) \wedge J \big] \\
&\quad - \frac{\alpha'}{2\kappa^2} \int_M \mathrm{dvol}_M \, e^{4A-2\Phi} \Big[2 e^{-2A} (\nabla^i \nabla^j e^A)(\nabla_i \nabla_j e^A) \\
&\quad \qquad + (\iota_j H \cdot \iota_j H) \nabla^i A \nabla^j A + 3 |\mathrm{d}A \cdot \mathrm{d}A|^2 \Big] \,. \qquad (4.1.6b)
\end{aligned}$$

Note that the potential (4.1.5) depends explicitly on the dilaton and fluxes, but both explicitly and implicitly on the metric, the latter through the SU(3) structure tensors J and Ω. Thus, in order to derive the equations of motion from this form of the potential, one needs to know the variations of J and Ω with respect to the metric, which are given by

$$\delta J = -\frac{1}{2}\delta g^{ij} \, g_{k(i}\mathrm{d}y^k \wedge \iota_{j)} J \,, \qquad \delta \Omega = -\frac{1}{2}\delta g^{ij} \, g_{k(i}\mathrm{d}y^k \wedge \iota_{j)} \Omega \,, \qquad (4.1.7)$$

---

[1] To obtain (4.1.3) from (C.1) of [197], one should set $f = 1$, $A = \Phi = H = 0$, $\Psi_1 = ie^{iJ}$ and $\Psi_2 = \Omega$ therein.
[2] We also denote the heterotic potential V_h by V in this chapter.

where δg^{ij} is a general variation of the inverse of the six-dimensional metric. Following the steps of section 2.4.2, would of course yield again that dA is zero up to order α'^3. However, we keep these terms for the moment for the sake of completeness.

In principle, one should also express the curvature R_+ in terms of the SU(3) structure forms J and Ω and the flux H, but this turns out to be not necessary for our purposes. One can use the decompositions in (p,q)-forms induced by the almost complex structure associated with J and Ω to rewrite the first two lines of the right hand side of (4.1.6b) as a sum of squares

$$\text{Tr}(F \wedge *F) + \text{Tr}(F \wedge F) \wedge J = \text{dvol}_M \left[2\,\text{Tr}\,|F^{(2,0)}|^2 + \text{Tr}(J\lrcorner F)^2 \right], \tag{4.1.8a}$$

$$\text{Tr}(R_+ \wedge *R_+) + \text{Tr}(R_+ \wedge R_+) \wedge J = \text{dvol}_M \left[2\,\text{Tr}\,|R_+^{(2,0)}|^2 + \text{Tr}(J\lrcorner R_+)^2 \right]. \tag{4.1.8b}$$

Note that by this scalar potential approach we have followed a philosophy quite similar to the one in [137], where a similar potential was constructed. Let us however point out a few differences between our potential and the one obtained there. First, we are not assuming constant warping. While this aspect will not be crucial for most of the discussions on compactifications with constant warping, allowing for a non-trivial warping makes explicit the consistency of our truncation ansatz as can be seen by our discussion of the EoM's in section 2.4.2. Secondly, our potential (4.1.5)-(4.1.6) is expressed in terms of the SU(3)-invariant (3,0)-form Ω, and not of the associated almost complex structure as in [137]. Finally, and most importantly, our potential is a sum of squares, while the potential of [137] is not, since it contains an $\mathcal{O}(\alpha'^0)$-term linear in the curvature. We will show next that all square terms vanish separately for non-broken SUSY, meaning that our potential is of BPS-like form. As we will see, having a fully-BPS structure will be crucial in studying possible mechanisms of supersymmetry breaking.

Supersymmetric vacua from the BPS potential

As explained in section 2.4.2, any vacuum must extremize the potential (4.1.5)-(4.1.6). Since this potential is a sum of squares, the simplest possibility is that each of these squares vanish separately. Let us first consider the $\mathcal{O}(\alpha'^0)$ potential V_0 (4.1.6a). Imposing that all squares vanish demands that the warping should be constant, $dA = 0$, in agreement with the discussion of section 2.4.2, and that the following equations should be satisfied:

$$d(e^{-2\Phi}\Omega) = 0, \tag{4.1.9a}$$

$$d(e^{-2\Phi}J \wedge J) = 0, \tag{4.1.9b}$$

$$e^{2\Phi}d(e^{-2\Phi}J) = *H. \tag{4.1.9c}$$

These match the conditions obtained in [130, 134] by standard spinorial arguments. Let us shortly review how to obtain them.

After our compactification to four dimensions one can deduce two equations from the gravitino variation (2.2.6a) by specifying to the external ($I = \mu$) or internal ($I = i$) components of the gravitino. By applying the spinor decomposition (4.1.1) and our gamma matrix conventions form appendix A.1 one finds for the external components

$$\delta\Psi_\mu = \frac{1}{2}e^A\hat{\gamma}_\mu\zeta \otimes (\slashed{\partial}A\eta_+ + e^{-A}w_0\eta_+^*) + \text{c.c.} = 0,$$
$$\Rightarrow \slashed{\partial}A\eta_+ + e^{-A}w_0\eta_+^* = 0. \quad (4.1.10)$$

Multiplying this equation by $(\eta_+^\dagger\gamma_i)$ and η_+^T, and using the properties of the SU(3) spinor η_+ discussed in section 3.2, one finds that $\mathrm{d}A$ and w_0 must be zero for a supersymmetric vacuum. This means that the $\mathcal{O}(\alpha'^3)$ result that we have obtained from the equations of motion has to hold to all orders in α' when SUSY is not broken.

Turning to the internal components of (2.2.6a) and the dilatino equation (2.2.6b) we find the two equations

$$\left(\nabla_m - \frac{1}{4}\slashed{H}_m\right)\eta_+ = 0, \quad (4.1.11a)$$
$$\left(\slashed{\partial}\phi - \frac{1}{2}\slashed{H}\right)\eta_+ = 0. \quad (4.1.11b)$$

These can be used to calculate the derivatives of J and Ω, which turn out to give exactly (4.1.9).[3] We see therefore that by imposing SUSY all squares in the order α'^0 contribution to the potential vanish. Therefore V_0 is extremized automatically in the supersymmetric case.

Note that in order to satisfy (4.1.9) the choices for the internal manifold M are quite restricted. In particular, (4.1.9a) requires M to be a complex manifold with a nowhere vanishing globally defined holomorphic $(3,0)$-form. The second condition (4.1.9b) requires the internal space to be conformally balanced [220]. Finally, the third condition (4.1.9c) imposes that, in presence of a non-vanishing three-form flux H the space is not Kähler.

By introducing the complexified three-form $\mathcal{G} = H - ie^{2\Phi}\mathrm{d}(e^{-2\Phi}J)$, one can see (4.1.9c) as an imaginary-self-duality (ISD) condition

$$*\mathcal{G} = i\mathcal{G}, \quad (4.1.12)$$

---

[3]For example

$$\mathrm{d}J = \frac{1}{2}\nabla_i J_{jk}\mathrm{d}x^{ijk} = \frac{1}{2\|\eta_+\|^2}\left((\nabla_i\|\eta_+\|^2)J_{jk} + i[(\nabla_i\eta_+)^\dagger\gamma_{jk}\eta_+ + \eta_+^\dagger\gamma_{jk}(\nabla_i\eta_+)]\right)\mathrm{d}x^{ijk}.$$

which means that $\mathcal{G}^{2,1}$ is primitive, $\mathcal{G}^{3,0} = 0$, and $\mathcal{G}^{1,2} = \eta \wedge J$ for some (0,1)-form η.[4]

The above supersymmetry equations can be rewritten in a slightly different form, by introducing another three-form
$$G = H - i\, dJ = \mathcal{G} - 2i\, d\Phi \wedge J\, . \tag{4.1.13}$$
Indeed, one can first use (4.1.9b) to rewrite (4.1.9c) as: $G^{3,0} = 0 = G^{1,2}$. Then, by noticing that (4.1.9a) implies via (3.2.10) that $(dJ)^{3,0} = 0$, (4.1.9) can be rewritten as

$$d(e^{-2\Phi}\Omega) = 0\, , \tag{4.1.14a}$$
$$d(e^{-2\Phi} J \wedge J) = 0\, , \tag{4.1.14b}$$
$$G^{1,2} = 0 = G^{0,3}\, . \tag{4.1.14c}$$

As we will see in section 4.4.2 expressing the supersymmetry conditions as in (4.1.14c) is more natural from the viewpoint of the effective four-dimensional theory. On the other hand, (4.1.9c) has a direct interpretation in terms of calibrations as discussed in section 4.3.

Let us finally rewrite (4.1.9) in terms of torsion classes (3.2.10). From (4.1.9a) and (4.1.9b) it is clear that
$$d\Phi = W_4 = \operatorname{Re} W_5\, , \tag{4.1.15}$$
and that $W_1 = W_2 = 0$. One can then deduce from (4.1.9c) that the flux H has no (3,0)- or (0,3)-components and that
$$H^{1,2} = i\, W_4^{0,1} \wedge J + i\, W_3^{1,2}\, . \tag{4.1.16}$$
From these considerations one can also see immediately that $G^{3,0}$, $G^{0,3}$, and $G^{1,2}$ have to be zero for supersymmetric vacua.

Next, we discuss the $\mathcal{O}(\alpha')$ potential (4.1.6b). Again, we require squared terms to vanish separately. The terms in the last line automatically vanish, since we have already imposed that A is constant. On the other hand, from the first line in (4.1.6b) and via (4.1.8a) one gets the conditions
$$F^{0,2} = 0\, , \quad J \lrcorner F = 0\, . \tag{4.1.17}$$
This means that the gauge bundle should be holomorphic, and should moreover have a primitive field-strength or, in other words, the gauge bundle must be Hermitian-Yang-Mills (HYM). That this is the case can be seen by plugging our compactification ansatz into the gaugino variation

---

[4]Note that a (3,0)-form is always imaginary anti-self-dual. A (1,2)-form is imaginary anti-self-dual if it is primitive. Hence, the (3,0)-part of $\mathcal{G}$ is zero and its (1,2)-part is non-primitive, respectively.

(2.2.6c). We find that

$$\not{F}\eta_+ = -i\,(J \cdot F)\,\eta_+ + \frac{1}{2}\,(\iota_m \Omega \cdot F)\,\gamma^m \eta_+^* = 0\,, \qquad (4.1.18)$$

which gives the conditions (4.1.17) when contracted with $\eta_+^\dagger$ and $\eta_+^T \gamma_i$.

Finally, from the second line of (4.1.6b) or (4.1.8b) one gets

$$R_+^{0,2} = 0\,, \quad J \lrcorner R_+ = 0 \qquad (4.1.19)$$

Conditions which, up to higher order α' corrections, are automatically implied by the supersymmetry conditions as was shown in [221]. Therefore, due to the fact that it is of BPS-like form, the whole potential (4.1.6) will be extremized automatically for a supersymmetric vacuum. In the next section we will analyze what can be learned from the potential if we drop one of the SUSY conditions (4.1.9).

4.2 Supersymmetry breaking vacua: general discussion

Let us now address the possible patterns of supersymmetry breaking in purely bosonic heterotic vacua. We will identify a particularly natural possibility, which we will further restrict in section 4.4 to a rather simple subfamily of constructions. As we will see in section 4.5 this restricted class of vacua include as a subcase the supersymmetry-breaking backgrounds considered in [190], which were mainly motivated by duality arguments.

4.2.1 Torsion induced SUSY-breaking vacua

Our strategy to construct non-supersymmetric vacua can be divided into two steps. First, we look for a supersymmetry breaking ansatz such that it violates the SUSY conditions of subsection 4.1, but still leads to a vanishing potential V, since this is still required by the equations of motion. As a second step, we need to consider whether V can be extremized within this ansatz, and which further constraints an extremization may impose.

Focusing on the $\mathcal{O}(\alpha'^0)$ piece of the potential, one sees that V_0 is the sum of positive and negative definite terms, and that the violation of the supersymmetry conditions implies that some of the positive definite terms do not vanish. Hence, $V_0 = 0$ is only possible if there is an exact cancellation between positive and negative definite terms in (4.1.6a).

From the general remarks of section 2.4.2, we know that the warping should be constant up to order α'^2, and so we can already set $dA = 0$. While we are still left with a large number of terms in V, a drastic simplification is obtained by imposing that the conditions (4.1.9c) and

(4.1.9b) are not violated in the non-supersymmetric vacuum. As we will discuss in section 4.3 this guarantees the geometrical structure needed in order to define the stability of the gauge bundle and of space-time filling NS5-branes. It is therefore natural to maintain these conditions in the context of compactification with a stable gauge sectors. To summarize, we impose

$$\mathrm{d}(e^{-2\Phi} J \wedge J) = 0, \quad (4.2.1\mathrm{a})$$

$$e^{2\Phi}\mathrm{d}(e^{-2\Phi} J) = *H \quad (\Leftrightarrow *\mathcal{G} = i\mathcal{G}), \quad (4.2.1\mathrm{b})$$

but we allow for

$$\mathrm{d}(e^{-2\Phi}\Omega) \neq 0. \quad (4.2.2)$$

This choice makes all the terms of V_0 containing derivatives of J vanish and encodes the origin of the supersymmetry breaking in the violation of (4.1.9a). If we further simplify the potential by imposing that

$$\overline{\Omega}\lrcorner \mathrm{d}(e^{-2\Phi}\Omega) = 0, \quad (4.2.3)$$

then the last line on the right hand side of (4.1.6a) also vanishes. We are thus left with the following non-vanishing contributions to the potential

$$V_0' = \frac{1}{4\kappa_{10}^2} \int \mathrm{dvol}_M \, e^{4A-2\Phi} \Big[|e^{-3A+2\Phi}\mathrm{d}(e^{3A-2\Phi}\Omega)|^2 - |J \wedge \mathrm{d}\Omega|^2 \Big], \quad (4.2.4)$$

which, taking into account that $\mathrm{d}A = 0$, vanishes if and only if [5]

$$|e^{2\Phi}\mathrm{d}(e^{-2\Phi}\Omega)|^2 = |J \wedge \mathrm{d}\Omega|^2. \quad (4.2.5)$$

Hence, one sees from the right hand side of (4.2.5) that it is the $(2,2)$-component of (4.2.2) which induces the SUSY-breaking

$$\text{SUSY-breaking} \quad \Leftrightarrow \quad \mathrm{d}(e^{-2\Phi}\Omega)^{2,2} \neq 0. \quad (4.2.6)$$

An important implication of this condition is that this supersymmetry breaking mechanism is possible only if the complex structure defined by the SU(3) structure is not integrable.

This way of breaking supersymmetry can be seen as the heterotic counterpart of the type II supersymmetry breaking pattern discussed in [197], which generalizes the flux-induced SUSY-

---

[5]Notice that we could have had vanishing potential even by violating the condition (4.2.1a), due to a non-trivial cancellation of the terms containing $\mathrm{d}(e^{2A-2\Phi} J \wedge J)$ in (4.1.6a). However, in this case the extremization of these terms in (4.1.6) is not straightforward and needs to be checked separately. This kind of SUSY-breaking can be thought as driven by D-terms and removes part of the integrable geometrical structure which could be crucial to study the stability of the bundle, cf. sections 4.3 and 4.4.2 below.

breaking pattern of type IIB warped Calabi-Yau/F-theory backgrounds [133]. In [197], this mechanism was named 'domain-wall supersymmetry breaking' (DWSB) because of its interpretation in terms of calibrations. As we will discuss in section 4.3, this interpretation is possible in the heterotic case as well, and so the present vacua will be named in the same manner.

In order to make contact with the flux literature, it is useful to translate the above conditions to the language of torsion classes (3.2.10). As in the supersymmetric case, (4.2.1a) and (4.2.3) are equivalent to fixing W_4 and W_5 in terms of the dilaton

$$W_4 = d\Phi, \qquad \text{Re}\, W_5 = d\Phi, \qquad (4.2.7)$$

while (4.2.3) and (4.2.5) give conditions on W_1 and W_2

$$e^{2\Phi} d(e^{-2\Phi}\Omega) = W_1 J \wedge J + W_2 \wedge J, \qquad \text{with} \qquad |W_2|^2 = 24|W_1|^2, \qquad (4.2.8)$$

where we recall that W_2 is a primitive $(1,1)$-form. Finally, (4.2.1b) can be rewritten as

$$H^{3,0} = -\frac{3}{4}\overline{W}_1 \Omega, \qquad (4.2.9a)$$

$$H^{2,1} = -i\,W_4^{1,0} \wedge J - i\,W_3^{2,1}. \qquad (4.2.9b)$$

Note that in this language the supersymmetry breaking can be associated to a non-vanishing W_1, that is, a non-vanishing $(dJ)^{3,0}$. As one can see from equation (4.2.9a) this is directly related to a non-vanishing $H^{3,0}$-component of the flux.

We can also characterize this supersymmetry breaking mechanism in terms of the three-form G. By inserting the expressions for H and $d\Phi$ into (4.1.13) one finds

$$G^{3,0} = G^{1,2} = 0 \qquad (4.2.10)$$

as in the supersymmetric case. The origin of supersymmetry breaking can then be traced back to the non-vanishing of $G^{0,3}$.

$$\text{SUSY-breaking} \quad \Leftrightarrow \quad G^{0,3} = -\frac{3}{2} W_1 \overline{\Omega} \neq 0. \qquad (4.2.11)$$

Formally, this condition is identical to the SUSY-breaking condition in type IIB warped Calabi-Yau/F-theory backgrounds [133], where G is constructed as $G = F_{\text{RR}} + ie^{-\Phi} H$ with F_{RR} the Ramond-Ramond three-form flux.

We also want to analyze how the conditions (4.2.1) – (4.2.3) alter the Killing spinor equations (4.1.11) and the external SUSY condition (4.1.10). In fact, due to our ansatz it could be that

$dA = 0 = w_0$ is no longer guaranteed by supersymmetry. However, since we know that in any vacuum these two quantities have to be zero up to $\mathcal{O}(\alpha'^3)$ in order to satisfy the equations of motion, we conclude that also (4.1.10) is not changed up to negligible α' corrections.

What can one say about (4.1.11)? Here, things are a bit more complicated. According to the chirality of η_+ one can parametrize the violation of $\delta \Psi_i = 0 = \delta \lambda$ by

$$\left(\nabla_i - \frac{1}{4}\slashed{H}_i\right)\eta_+ = p_i\,\eta_+ + q_{ij}\gamma^j \eta_+^*\,, \tag{4.2.12a}$$

$$\left(\slashed{\partial}\phi - \frac{1}{2}\slashed{H}\right)\eta_+ = u_i \gamma^i \eta_+ + r\,\eta_+^*\,. \tag{4.2.12b}$$

One should note here that due to the contraction with one gamma matrix the second index of q_{ij} is of $(1,0)$-type, while u_i is a $(0,1)$-form[6]

$$P_i{}^k q_{jk} = q_{ji}\,, \qquad \overline{P}_i{}^k u_k = u_i\,. \tag{4.2.13}$$

Note also that q_{ij} is not antisymmetric in its indices. Using this parametrization one can again compute the derivatives of J and Ω which turn out to be

$$\begin{aligned}
e^{2\phi}\mathrm{d}(e^{-2\phi}J \wedge J) &= 4\operatorname{Re}(p-u)\wedge J \wedge J - 8\operatorname{Re}(\bar{s}\wedge\Omega)\,, & (4.2.14a)\\
e^{2\phi}\mathrm{d}(e^{-2\phi}J) - *H &= 2\operatorname{Im}(r^*\Omega) + 2\operatorname{Re}(p-2u)\wedge J - 2\operatorname{Im}(t^{*i}\wedge \iota_i\Omega)\,, & (4.2.14b)\\
e^{2\phi}\mathrm{d}(e^{-2\phi}\Omega) &= 2(p-u)\wedge\Omega - rJ\wedge J + 8is\wedge J\,. & (4.2.14c)
\end{aligned}$$

Here, we defined $s = \frac{1}{2}q_{ij}\,\mathrm{d}y^i \wedge \mathrm{d}y^j$, $t^j = q_i{}^j\mathrm{d}y^i$, $u = u_i\mathrm{d}y^i$ and $p = p_i\mathrm{d}y^i$. Comparing these equations to our ansatz (4.2.1) – (4.2.3) one sees that the first two of them must be equal to zero. In the first equation both terms on the right hand side have to vanish separately. This gives the conditions

$$\operatorname{Re} u = \operatorname{Re} p\,, \qquad \bar{s}\wedge\Omega = 0\,, \qquad s\wedge\overline{\Omega} = 0\,, \tag{4.2.15}$$

which implies that s is a $(1,1)$-form. Then, the last term on the right hand side of the second equation is $(3,0) + (0,3)$. This means that by contracting this equation with $\overline{\Omega}$ one can relate r to the trace of q_{ij}

$$r = q_{ij}\,g^{ij}\,. \tag{4.2.16}$$

Furthermore, the middle term has to vanish separately, leading to

$$\operatorname{Re}(p - 2u) = -3\operatorname{Re} u = -3\operatorname{Re} p = 0\,. \tag{4.2.17}$$

---
[6]This can be seen by e.g. multiplying (4.2.12b) from the left by $\eta_+^T \gamma^{jk}$. Then one obtains a term of the form $u_i\Omega^{ijk}$. Since Ω is a $(3,0)$-form and raising an index changes it form holomorphic to antiholomorphic (and vice versa), this means that u_i is of $(0,1)$-type.

Since u is a $(0,1)$-form the vanishing of $\mathrm{Re}\,u$ means that $u = 0$

$$\mathrm{Re}\,u \;=\; \tfrac{1}{2}\mathrm{Re}\,(u + iJ\lrcorner u) \;=\; \tfrac{1}{2}(\mathrm{Re}\,u - J\lrcorner\mathrm{Im}\,u) \;=\; 0\,,\quad \Rightarrow\quad u = 0\,. \tag{4.2.18}$$

At last, contracting the third equation of (4.2.14) with $\overline{\Omega}$ and demanding (4.2.3) sets also $\mathrm{Im}\,p = 0$. Thus, we see that the SUSY-breaking condition (4.2.2) takes the form

$$e^{2\phi}\mathrm{d}\!\left(e^{-2\phi}\Omega\right) \;=\; -r\,J\wedge J + 8\,i\,s\wedge J\,, \tag{4.2.19}$$

where s is a $(1,1)$-form and can be decomposed in a primitive and a non-primitive part $s = -\tfrac{i}{6}r\,J + s_{\mathrm{P}}$. Comparing this with the first condition in (4.2.8), we find $r = 3W_1$ and

$$s \;=\; -\frac{i}{8}\,(W_2 + 4W_1 J)\,. \tag{4.2.20}$$

From this one can finally see that the violations of the gravitino and dilatino Killing spinor equations are of the form

$$\delta\Psi_\mu \;=\; 0\,, \tag{4.2.21a}$$

$$\delta\Psi_i \;=\; -\frac{i}{4}\mathcal{S}_{ij}\,\zeta\otimes\gamma^j\eta_+^* + \text{c.c.}\,, \tag{4.2.21b}$$

$$\delta\lambda \;=\; 3W_1\,\zeta\otimes\eta_+^* + \text{c.c.}\,, \tag{4.2.21c}$$

where we have introduced the two-form[7]

$$\mathcal{S} \;=\; W_2 + 4W_1 J\,. \tag{4.2.22}$$

In order to conclude our $\mathcal{O}(\alpha'^0)$ discussion, it remains to impose the extremization of V_0, which unlike the supersymmetric case is not automatic. However, it is sufficient to impose the extremization of V_0' given in (4.2.4), since all other terms are automatically extremized, being quadratic in vanishing terms. In particular, it is easy to see that the only non-trivial contribution comes from the extremization of V_0' under variations of the metric. By using equation (4.1.7), the resulting residual equations of motion are given by

$$\mathrm{Im}\!\left[\iota_{(i}\overline{\Omega}\lrcorner\iota_{j)}\mathrm{d}\mathcal{S}\right] \;=\; 8\,g_{ij}|W_1|^2 - 2\mathrm{Re}[\overline{W}_1(\iota_i W_2\lrcorner\iota_j J)] - \mathrm{Re}[\iota_i W_2\lrcorner\iota_j \overline{W}_2]$$

---
[7]Note that there is a relative factor of 2 between $(-i/4\mathcal{S})$ and s. This is because $\mathcal{S}$ is a generic $(1,1)$-form, whose second index could be holomorphic as well as anitholomorphic, while the second index of s can only be holomorphic, due to its connection to q.

$$= |W_1|^2 \Big\{ 9\, g_{ij} - \mathrm{Re}\Big[\iota_i\Big(\frac{W_2}{W_1}+J\Big)\lrcorner\iota_j\Big(\frac{\overline{W}_2}{\overline{W}_1}+J\Big)\Big]\Big\}\,. \qquad (4.2.23)$$

Due to the contraction with the $(0,3)$-form $\overline{\Omega}$, only the $(3,0)$ and primitive $(2,1)$ components of $\mathrm{d}\mathcal{S}$ can contribute to the left hand side of (4.2.23). Since the right hand side is a $(1,1)$-tensor, it can only be matched to the left hand side if $\mathrm{d}\mathcal{S}$ has only $(3,0)$ components. This implies that the primitive $(2,1)$ components of $\mathrm{d}\mathcal{S}$ must vanish

$$(\mathrm{d}\mathcal{S})_{\mathrm{P}}^{(2,1)} = 0\,. \qquad (4.2.24)$$

In fact, the $(3,0)$ component of $\mathrm{d}\mathcal{S}$ also vanishes, as can be seen by using (4.2.22) and (4.2.8).[8] We conclude that the right hand side of (4.2.23) must vanish identically. By introducing the matrix

$$U^i{}_j = \frac{1}{3}\Big(\frac{W_2}{W_1}+J\Big)^i{}_j = \frac{1}{3}g^{ik}\Big(\frac{W_2}{W_1}+J\Big)_{kj}\,, \qquad (4.2.25)$$

we arrive at the following matrix equation

$$\mathrm{Re}\,(U^i{}_k \overline{U}_j{}^k) = \mathrm{Re}\,(U \cdot U^\dagger) = \mathbb{1}\,, \qquad (4.2.26)$$

where $(U^\dagger)^i{}_j = (\overline{U}_j{}^i) = g_{jk}g^{il}(\overline{U}^k{}_l)$. To sum up, the equations of motion (4.2.23) boil down to the conditions (4.2.24) and (4.2.26). Furthermore, the primitivity of W_2 implies that

$$J^i{}_j U^j{}_i = \mathrm{Tr}(JU) = -2\,. \qquad (4.2.27)$$

While at this point these conditions look rather mysterious, we will provide a simple geometrical interpretation for them in section 4.4.

At order α' we also need to impose the extremization of the term V_1 in (4.1.6b). Since $\mathrm{d}A = 0$, terms containing the warping do not provide further constraints. On the other hand, the terms containing the gauge bundle field-strength F are extremized if

$$F^{0,2} = 0\,, \quad J\lrcorner F = 0 \qquad (4.2.28)$$

as in the supersymmetric case. Recall that the almost-complex structure of M is not integrable. Following [222], we can say that (4.2.28) requires the bundle to be pseudo-HYM and in particular that the condition $F^{0,2}=0$ requires the bundle to be pseudo-holomorphic. Clearly, one cannot use the standard theory of bundles on Kähler spaces. However, as we will discuss in more detail in section 4.3, the conditions (4.2.1a) and (4.2.1b) still allow to define a sort of stability of the gauge

---

[8]This component is proportional to $\mathrm{d}\mathcal{S} \wedge \overline{\Omega} = -\mathcal{S} \wedge \mathrm{d}\overline{\Omega}$, which vanishes upon imposing (4.2.8).

bundle, analogous to the one for bundles on Kähler spaces.

Finally, from the second line of (4.1.6b) one gets

$$R_+^{0,2} = 0\,, \quad J \lrcorner R_+ = 0\,. \tag{4.2.29}$$

Unlike the supersymmetric case, these equations are not any longer automatically satisfied. However, imposing that the supersymmetry breaking is mild enough compared to the compactification scale, the violation of (4.2.29) is expected to be mild as well, and possibly negligible at $\mathcal{O}(\alpha')$. We will come back to this point in subsection 4.2.3, but focus now on how our SUSY-breaking ansatz affects gravitino and gaugino masses.

4.2.2 Gravitino and gaugino mass

While equations (4.2.6) and (4.2.11) translate the fact that supersymmetry is broken into geometry, one can also provide a more physical measure of the amount of SUSY-breaking by computing the four-dimensional gravitino and gaugino masses. Note that in general a simple consistent truncation ansatz does not necessarily exists for these backgrounds and, as a result, there is no precise definition of the four-dimensional gravitino. Nevertheless, one can introduce a sort of a four-dimensional gravitino Ψ_μ^{4D} defined by the fermionic decomposition

$$\Psi_\mu + \frac{1}{2}\Gamma_\mu(\Gamma^m \Psi_m - \lambda) \;=\; \Psi_\mu^{4D} \otimes \eta_+^* + \text{c.c.}\,. \tag{4.2.30}$$

Note that defined in this way Ψ_μ^{4D} may depend on the internal coordinates, and thus cannot be considered as a properly defined four-dimensional gravitino. Nevertheless, the combination of the ten-dimensional gravitino and dilatino appearing on the left hand side of (4.2.30) is such that the four-dimensional kinetic term for Ψ_μ^{4D}, resulting from the ten-dimensional action of [200][9] has a canonical form and does not mix with other fermions.

One can now introduce a function $\mathfrak{m}_{3/2}$ which plays the role of the gravitino mass but generically also depends on the internal coordinates. Indeed, let us define $\mathfrak{m}_{3/2}$ as follows

$$\delta \Psi_\mu^{4D} \;=\; \frac{1}{2}\mathfrak{m}_{3/2}\,\hat\gamma_\mu \zeta\,, \tag{4.2.31}$$

i.e. by using the usual four-dimensional SUSY-breaking formula which relates the variation of the gravitino to the gravitino mass.[10]

---

[9]Our conventions and the ones used in [200] are related by: $\Phi^{\text{there}} = \exp(2\Phi^{\text{here}}/3)$, $H^{\text{there}} = H^{\text{here}}/3\sqrt{2}$, $\Psi_M^{\text{there}} = \Psi_M^{\text{here}}$, $\lambda^{\text{there}} = -\lambda^{\text{here}}/2\sqrt{2}$, and $\chi^{\text{there}} = -\chi^{\text{here}}$.

[10]In order to identify the four-dimensional spinor ζ in (4.1.1) with the generator of the four-dimensional supersymmetry, it is convenient to choose the normalization $\eta_+^\dagger \eta_+ = e^A$ for the internal spinor η_+. This is possible since

Applied to the SUSY-breaking backgrounds described in subsection 4.2.1, by (4.2.21) we see that the above definitions yield the expression

$$\mathfrak{m}_{3/2} = 3 e^A W_1 \,. \tag{4.2.32}$$

So $\mathfrak{m}_{3/2}$ can be related to the scale set by the (4d normalized) torsion class W_1. Note again that this scale depends on the coordinates of the internal manifold.

In order to make contact with a four-dimensional effective theory, one would like to have a more standard expression for the gravitino mass $m_{3/2}$. This can be obtained by imposing Ψ_μ^{4D} to be constant in the internal space and averaging $\mathfrak{m}_{3/2}$ with an appropriate dilaton-factor

$$m_{3/2} = \langle \mathfrak{m}_{3/2} \rangle = \frac{\int_M \mathrm{dvol}_M \, e^{-2\Phi} \mathfrak{m}_{3/2}}{\int_M \mathrm{dvol}_M \, e^{-2\Phi}} = \frac{i e^A \int_M e^{-2\Phi} \, \Omega \wedge G}{4 \int_M \mathrm{dvol}_M \, e^{-2\Phi}}, \tag{4.2.33}$$

where in the last step we have used (4.2.11) and the condition $G^{3,0} = 0$ for the three-form G defined in (4.1.13). One can then use (2.4.34) to fix the four-dimensional Einstein frame and gets

$$m_{3/2} = \frac{i \, g_s^3 l_s^4 \, M_\mathrm{P} \int_M e^{-2\Phi} \, \Omega \wedge G}{8 \sqrt{\pi} \, \mathrm{Vol}(M)^{3/2}} \,. \tag{4.2.34}$$

Let us now turn to the gaugino mass. The four-dimensional gaugino χ^{4D} is related to the ten-dimensional gaugino by the decomposition

$$\chi = e^{-2A} \chi^{4D} \otimes \eta_+ + \mathrm{c.c.} \,. \tag{4.2.35}$$

The relevant terms in the ten-dimensional action [200] are given by

$$-\frac{\alpha'}{4\kappa^2} \int \mathrm{d}^{10}x \sqrt{-g} \, e^{-2\Phi} \left(\mathrm{Tr} \, \overline{\chi} \slashed{\nabla} \chi - \frac{1}{4} \mathrm{Tr} \, \overline{\chi} \slashed{H} \chi \right), \tag{4.2.36}$$

Plugging (4.2.35) into (4.2.36) and integrating over the internal space, while keeping χ^{4D} constant on it, one obtains the following value for the gaugino mass

$$m_{1/2} = \frac{i e^A \int_M e^{-2\Phi} \, \Omega \wedge H}{2 \int_M \mathrm{dvol}_M \, e^{-2\Phi}} = \frac{i e^A \int_M e^{-2\Phi} \, \Omega \wedge G}{4 \int_M \mathrm{dvol}_M \, e^{-2\Phi}}, \tag{4.2.37}$$

where in the last step we have used the condition $G^{3,0} = 0$ again. We thus see that the gaugino

---

Re $p = 0$ which implies $\nabla_i \|\eta_+\|^2 = 0$.

mass equals the gravitino mass (4.2.33)

$$m_{1/2} = m_{3/2}. \qquad (4.2.38)$$

As we will see in section 4.4.2, this result has a very simple four-dimensional interpretation.

4.2.3 Conditions on the curvature

Let us now discuss the conditions on the curvature (4.2.29) that arise at order α' from the minimization of the potential piece (4.1.6b). First of all, note that

$$R_{+ijkl} = R_{-klij} - (\mathrm{d}H)_{ijkl}. \qquad (4.2.39)$$

So, by using the BI (2.2.5) we get the relation

$$R_{+ijkl} = R_{-klij} + \mathcal{O}(\alpha'). \qquad (4.2.40)$$

Hence, in the scalar potential the terms (4.2.29) can be rewritten as

$$\Omega_{ijk} R_{-}^{ij} = 0, \qquad J_{ij} R_{-}^{ij} = 0 \qquad (4.2.41)$$

up to $\mathcal{O}(\alpha'^2)$. These conditions can be rephrased by saying that the internal spinor η_+ specifying the SU(3) structure should be covariantly constant with respect to the torsion-full covariant derivative ∇_i^-. From (4.2.12a) and (4.2.21b), we know that this is not the case in the torsional SUSY-breaking backgrounds of subsection 4.2.1. However, let us assume that the SUSY-breaking is mild, so that $\nabla_i^- \eta_+ \sim \mathcal{O}(\alpha'^\beta)$, with $0 < \beta \leq 1$. Roughly speaking, this would mean that both equations in (4.2.29) are violated at $\mathcal{O}(\alpha'^\beta)$. In particular, by using (4.1.8b), the curvature squared term in (4.1.6b) would be of $\mathcal{O}(\alpha'^{2\beta})$, and so negligible in our approximation for $\beta \geq 1/2$. Under this condition, the full potential would be extremized at our level of accuracy.

We can make this argument more concrete. From (4.2.12a) and (4.2.21b) we have

$$\nabla_i^- \eta_+ = -\frac{i}{4} \mathcal{S}_{ij} \gamma^j \eta_+^*. \qquad (4.2.42)$$

Taking into account (4.2.22) and the condition $|W_2|^2 = 24|W_1|^2$, we get qualitatively $\nabla_i^- \eta_+ \sim W_1 \gamma_i \eta_+^*$. The torsion class W_1 has the dimension of mass and defines a dimensionless SUSY-breaking length scale L_{SB} (measured in string units) through

$$W_1 \sim (l_s L_{\mathrm{SB}})^{-1} \qquad (4.2.43)$$

Then, taking $g_{ij} \sim l_s^2 L_{KK}^2$, with L_{KK} being the KK length measured in string units we have $\nabla_i^- \eta_+ \sim L_{KK} L_{SB}^{-1}$. Furthermore, by introducing the four-dimensional KK-scale $M_{KK} = e^A/(l_s L_{KK})$ and recalling (4.2.32), we can restate (4.2.43) in a more physical way

$$\mathfrak{m}_{3/2} \sim M_{KK} L_{KK} L_{SB}^{-1} \,. \qquad (4.2.44)$$

One has mild SUSY-breaking, which can be seen as spontaneous from the four-dimensional point of view, when $\mathfrak{m}_{3/2} \ll M_{KK}$. This condition corresponds to

$$\frac{L_{KK}}{L_{SB}} \ll 1 \,. \qquad (4.2.45)$$

This also means that the violation of SUSY in terms of spinors is small $\nabla_i^- \eta_+ \ll 1$.

Lets try to parametrize this relation in terms of L_{KK}. In the regime of the validity of a supergravity approximation all compactification scales should be larger than the string length which is guaranteed if

$$L_{KK} > 1 = \frac{\alpha'}{4\pi^2 l_s^2} = \frac{4\pi^2 l_s^2}{\alpha'} \,. \qquad (4.2.46)$$

Demanding that $\nabla_i^- \eta_+$ is of order α'^β (measured in string length, as $\nabla_i^- \eta_+$ is dimensionless) yields by (4.2.46)

$$\nabla_i^- \eta_+ \in \left(L_{KK}^{-2\beta}, L_{KK}^{2\beta} \right) \,. \qquad (4.2.47)$$

For minimal SUSY-breaking one should choose $\nabla_i^- \eta_+ \sim L_{KK}^{-2\beta}$, which is even smaller then $\mathcal{O}(\alpha'^\beta)$. Together with our former estimate $\nabla_i^- \eta_+ \sim L_{KK} L_{SB}^{-1}$ we get a relation between the KK length and the SUSY-breaking length

$$L_{SB} \sim L_{KK}^{2\beta+1} \,. \qquad (4.2.48)$$

If for example $\beta = 1/2$ in (4.2.48) we can identify L_{SB} with L_{KK}^2.

We can consider these issues also in a bit more detail. The curvature terms (4.2.41) can be rewritten by using (4.2.42) and the formula

$$[\nabla_i^-, \nabla_j^-] \eta_+ = \frac{1}{4} R_{-klij} \gamma^{kl} \eta_+ \,. \qquad (4.2.49)$$

They read

$$J_{i_1 i_2} R_-^{i_1 i_2}{}_{j_1 j_2} = 2 P^{lk} \mathcal{S}_{k[j_1} \mathcal{S}_{j_2]l}^* \,, \qquad (4.2.50a)$$

$$\Omega_{j_1 i_1 i_2} R_-^{i_1 i_2}{}_{j_2 j_3} = 4 i P_{j_1}{}^k \nabla_{[j_2}^- \mathcal{S}_{j_3]k} + \frac{1}{2} \Omega_{j_1}{}^{kl} \mathcal{S}_{k[j_2} \mathcal{S}_{j_3]l}^* \,, \qquad (4.2.50b)$$

where $P_i{}^j$ is defined in (3.2.5) and projects onto holomorphic indices of the almost complex struc-

64

ture. Then, we have the following curvature squared terms contributing to (4.1.6b)

$$|J_{ij}R_-^{ij}|^2 \sim |W_1|^4, \tag{4.2.51a}$$

$$|\Omega_{ijk}R_-^{ij}|^2 \sim |\partial W_1|^2 + |W_1^2\partial W_1| + |W_1|^4$$

$$\sim (l_s\, L_{\mathrm{KK}})^{-2}|W_1|^2 + (l_s\, L_{\mathrm{KK}})^{-1}|W_1|^3 + |W_1|^4\ . \tag{4.2.51b}$$

By using (4.2.43), the dimensionless contribution to the $\mathcal{O}(\alpha')$ equations of motions associated to the curvature terms in (4.1.6b) can be approximated as

$$(\mathrm{EoM})_{\mathcal{O}(\alpha')} \sim g_{ij}\, l_s^2 \left(|\Omega\lrcorner R_-|^2 + |J\lrcorner R_-|^2\right)$$

$$\sim \frac{1}{L_{\mathrm{KK}}^2}\left[\left(\frac{L_{\mathrm{KK}}}{L_{\mathrm{SB}}}\right)^2 + \left(\frac{L_{\mathrm{KK}}}{L_{\mathrm{SB}}}\right)^3 + \left(\frac{L_{\mathrm{KK}}}{L_{\mathrm{SB}}}\right)^4\right]\ . \tag{4.2.52}$$

Note that from g_{ij} one gets an extra factor of $l_s^2 L_{\mathrm{KK}}^2$. We have separated an overall factor of L_{KK}^{-2}, which gives a leading factor of $\mathcal{O}(\alpha')$, while the terms in squared brackets provides a further suppression because of (4.2.45). In order to make the correction (4.2.52) of order α'^2, for example, and therefore safely negligible at our $\mathcal{O}(\alpha')$ approximation, (4.2.52) should scale at least like L_{KK}^{-4}. Then, one would have to demand $L_{\mathrm{SB}} = L_{\mathrm{KK}}^2$, and thus $\beta = 1/2$. On the other hand, one could further relax this condition, depending on the details of the background. If for example in (4.2.51b) one finds $|\partial W_1| \lesssim |W_1|^2$, then it is enough to take $L_{\mathrm{SB}} = L_{\mathrm{KK}}^{3/2}$, i.e. $\beta = 1/4$.

We conclude that the curvature sector of the $\mathcal{O}(\alpha')$-correction restricts possible solutions such that only mild supersymmetry breaking is allowed. The mass scale of the SUSY-breaking has to be well below the compactification scale and hence the breaking of supersymmetry can be regarded as spontaneous from the four-dimensional perspective. The more severe restriction that we encountered in this section arose from the order α'^0 part of the potential. We will analyze possible solutions in section 4.4 and 4.5. But before we come to this, we will present another view of DWSB coming from the perspective of calibrations.

4.3 NS5-branes, calibrations and bundle stability

As already discussed in the literature (see e.g. [158, 159]) the supersymmetry conditions (4.1.9) admit a clear interpretation in terms of so-called p-form calibrations [191–193], which are p-forms that measure the energy of extended BPS objects of the theory. Classes of BPS objects are most conveniently separated in terms of their four-dimensional appearance, as illustrated in figure 4.1, since a different calibration exists for each four-dimensional BPS object. In particular, the two-form $e^{-2\Phi}J$ is a calibration for an NS5-brane that wraps an internal two-cycle of M and fills the four-dimensional space-time X_4. The four-form $e^{-2\Phi}J \wedge J$, on the other hand, is a calibration for NS5-branes wrapping an internal four-cycle and filling two directions in X_4 (showing up as 4d strings upon dimensional reduction). Finally, for any constant phase $e^{i\vartheta}$, $e^{-2\Phi}\text{Im}(e^{i\vartheta}\Omega)$ calibrates NS5-branes wrapped on three internal and three external directions, thus appearing as a domain-wall in four dimensions. More schematically, we have the following dictionary between calibrations and BPS objects of the compactification.[11]

Calibration	10d BPS object	4d BPS object
$e^{-2\Phi}J$	NS5 on $X_4 \times \Pi_2$	gauge theory
$e^{-2\Phi}\Omega$	NS5 on $X_3 \times \Pi_3$	domain wall
$e^{-2\Phi}J \wedge J$	NS5 on $X_2 \times \Pi_4$	string

Here, Π_p is a p-dimensional submanifold of M, and X_d is a d-dimensional slice of X_4. More precisely, the statement is that all these p-forms can be defined as calibrations only if the corresponding differential SUSY conditions (4.1.9) are satisfied. Now, recall that our SUSY-breaking pattern is characterized by (4.2.1) and (4.2.2). Hence, we see that, even if supersymmetry is broken, $e^{-2\Phi}J$ and $e^{-2\Phi}J \wedge J$ can still be identified as calibrations, while $e^{-2\Phi}\text{Im}(e^{i\vartheta}\Omega)$ cannot.

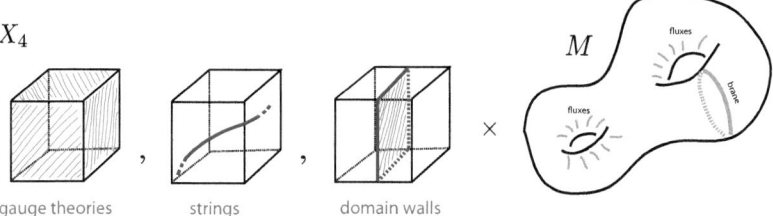

Figure 4.1: BPS objects of the theory, in terms of their four-dimensional appearance.

For this reason, we call this pattern 'domain-wall supersymmetry breaking' (DWSB) as done in [197] for the analogous case in the context of type II flux compactifications, which used the interpretation in terms of calibration provided by [194, 196, 223]. In order to understand better the implications of this observation, let us recall the main properties of a calibration.

In general, a calibration structure provides a natural BPS bound for certain branes in the ten-dimensional theory. Let us, for instance, consider NS5-branes filling X_4 and wrapping an internal two-cycle $\Pi_2 \subset M$. These branes couple magnetically to the three-form flux H and thus modify

---
[11]Following the discussion below, gauge bundles could also be added to this dictionary on the same footing as space-time filling NS5-branes.

the (internal) BI as

$$\mathrm{d}H = \frac{\alpha'}{4}\left(\mathrm{Tr}\, R_+ \wedge R_+ - \mathrm{Tr}\, F \wedge F\right) + 2\kappa^2\, \tau_{\mathrm{NS5}}\, \delta^4(\Pi_2)\,, \qquad (4.3.1)$$

where $\tau_{\mathrm{NS5}} = (2\pi)^{-5}(\alpha')^{-3}$ and $\delta^4(\Pi_2)$ is a four-form that localizes the NS5-brane in the four-dimensional space orthogonal to its world volume. It is clear that H and Π_2 cannot be considered as independent and for this reason it is convenient to go to a dual description, where the NS5-brane couples electrically to the seven-form flux $\hat{H} = \mathrm{dvol}_M \wedge (e^{-2\Phi} * H)$.[12] We can then write $\hat{H} = \mathrm{vol}_M \wedge \tilde{H}$, where $\tilde{H} = e^{-2\Phi} * H$ is a three-form of M. The dualization procedure in absence of NS5-branes is reviewed in appendix A.2 and leads to a dual formulation of the supergravity potential V introduced in section 4.2. The only modification due to the addition of an NS5-brane is the addition of the NS5-brane potential to V

$$V_{\mathrm{NS5}} = \tau_{\mathrm{NS5}} \int_{\Pi_2} \left(e^{-2\Phi}\sqrt{\det g|_{\Pi_2}}\,\mathrm{d}^2\sigma - \tilde{B}\right), \qquad (4.3.2)$$

where $\tilde{B}$ is the two-form potential of $\tilde{H}$, $\tilde{H} = \mathrm{d}\tilde{B}$, and $\mathrm{d}^2\sigma = \mathrm{d}\sigma^1 \wedge \mathrm{d}\sigma^2$ is the volume density induced by the world-volume coordinates (σ^1, σ^2) on Π_2.

As stated above, such a brane has the corresponding calibration $e^{-2\Phi} J$. This two-form provides the following algebraic inequality

$$e^{-2\Phi}\sqrt{\det g|_{\Pi_2}}\,\mathrm{d}^2\sigma \geq e^{-2\Phi} J|_{\Pi_2}\,, \qquad (4.3.3)$$

for any (appropriately oriented) Π_2. When the above inequality is saturated at every point of Π_2, then one says that the cycle Π_2 is calibrated:

$$\Pi_2 \text{ calibrated} \quad \Leftrightarrow \quad e^{-2\Phi}\sqrt{\det g|_{\Pi_2}}\,\mathrm{d}^2\sigma = e^{-2\Phi} J|_{\Pi_2}\,. \qquad (4.3.4)$$

Now, the differential condition (4.2.1b) can be rewritten as

$$\mathrm{d}(e^{-2\Phi} J) = \tilde{H}\,, \qquad (4.3.5)$$

and allows to prove the following important statement: an NS5-brane wrapping a calibrated two-cycle Π_2 globally minimizes its potential energy (4.3.2) under continuous deformations. More precisely, considering any other two-cycle Π_2' connected to Π_2 by a three chain Γ, $\partial\Gamma = \Pi_2' - \Pi_2$,

---

[12]In order to simplify the notation, in the remainder of this section we will set $e^A = 1$.

one gets $V_{\text{NS5}}(\Pi_2') \geq V_{\text{NS5}}(\Pi_2)$

$$V_{\text{NS5}}(\Pi_2') \geq \tau_{\text{NS5}} \int_{\Pi_2'} (e^{-2\Phi} J - \tilde{B}) = \tau_{\text{NS5}} \int_{\Pi_2} (e^{-2\Phi} J - \tilde{B}) = V_{\text{NS5}}(\Pi_2) \,. \quad (4.3.6)$$

In this inequality we have used (4.3.3) in the first step, the differential condition (4.3.5) in the second one, and the definition of calibrated cycles (4.3.4) in the last step, respectively. The same arguments equally apply for the calibration $e^{-2\Phi} J \wedge J$.

Thus, we see that in DWSB compactifications we have a natural notion of BPSness and stability for space-filling and string-like NS5-branes. These structures are typically associated to supersymmetric settings, and one can see them as a distinguished property of the SUSY-breaking pattern considered here, in analogy with the type II setting of [197].

In fact, the calibration structures provided by $e^{-2\Phi} J$ and $e^{-2\Phi} J \wedge J$ have also implications on the notion of gauge-bundle stability in the non-supersymmetric context. Since to the gauge bundle an induced NS5-brane charge density proportional to $\text{Tr}(F \wedge F)$ is associated (which is at the origin of the BI identity (4.3.1)), it is again convenient to work in the dual formulation reviewed in appendix A.2. In this formulation, one can isolate the following contributions of the gauge bundle to the total potential

$$V_{\text{bundle}} = \frac{1}{8\kappa^2} \int_M \left[e^{-2\Phi} \text{Tr}(F \wedge *F) + \tilde{B} \wedge \text{Tr}(F \wedge F) \right] \,. \quad (4.3.7)$$

Now, the bundle-analog of the inequality (4.3.3) for NS5-branes is

$$\text{Tr}(F \wedge *F) \geq -\text{Tr}(F \wedge F) \wedge J \,, \quad (4.3.8)$$

and the analog of the calibration condition (4.3.4) is the pseudo-HYM [222] condition[13]

$$\text{Tr}(F \wedge *F) = -\text{Tr}(F \wedge F) \wedge J \quad \Leftrightarrow \quad \begin{cases} F^{0,2} = 0 & \text{(F-flatness)} \\ J \wedge J \wedge \text{Tr}\, F = 0 & \text{(D-flatness)} \end{cases} \,. \quad (4.3.9)$$

Now, as for NS5-branes, one can easily show that the pseudo-HYM gauge bundles are absolute minima of V_{bundle} under continuous deformations, and again the differential condition (4.3.5) is crucial for the result. Indeed, suppose that F is a pseudo-HYM field strength and F' is any other field-strength which is cohomologous to F, so that there is a three-form α such that $\text{Tr}(F' \wedge F') = \text{Tr}(F \wedge F) + \text{d}\alpha$. Then, we have $V_{\text{bundle}}(F') \geq V_{\text{bundle}}(F)$, since

$$V_{\text{bundle}}(F') \geq -\frac{1}{8\kappa^2} \int_M \text{Tr}(F' \wedge F') \wedge (e^{-2\Phi} J - \tilde{B})$$

---
[13]The prefix 'pseudo' comes from the non-integrability of the almost-complex structure of M.

$$= -\frac{1}{8\kappa^2} \int_M \text{Tr}(F \wedge F) \wedge (e^{-2\Phi} J - \tilde{B}) = V_{\text{bundle}}(F), \qquad (4.3.10)$$

where we have used (4.3.8) in the first step, the differential condition (4.3.5) in the second one, and the pseudo-HYM condition (4.3.9) in the last one. As written in (4.3.9) the pseudo-HYM condition can be split into two parts. The first, $F^{0,2} = 0$, demands the bundle to be pseudo-holomorphic [222] and can be seen as an F-flatness condition, while the second one can be seen as a D-flatness condition [224].

Now, suppose that we can solve the F-flatness condition $F^{0,2} = 0$. Then, clearly the D-flatness in (4.3.9) admits a solution only if

$$\int_M e^{-2\Phi} J \wedge J \wedge \text{Tr}\, F = 0. \qquad (4.3.11)$$

This necessary condition is quasi-topological since it depends only on the first Chern class of the bundle, but changes under the deformations of $e^{-2\Phi} J \wedge J$. A natural question is then if one can extend this necessary condition in order to get a quasi-topological necessary and sufficient condition. In other words, one can ask whether a notion of quasi-topological stability exist for pseudo-holomorphic bundles in our backgrounds, which are only almost-complex spaces, but nevertheless admit the calibration structures described above.

In the complex (supersymmetric) case, the existence of a solution of the D-flatness equation for a holomorphic gauge bundle is equivalent to the so-called μ-stability of the bundle by the theorems of Donaldson-Uhlenbeck-Yau [225, 226], which are valid for Kähler spaces, and their generalizations to non-Kähler hermitian spaces [227]. In particular, in the non-Kähler case of interest for supersymmetric heterotic flux compactifications, the μ-stability of a bundle E is defined in terms of the μ-slope of E

$$\mu(E) = \frac{1}{\text{rank}\, E} \int_M e^{-2\Phi} J \wedge J \wedge \text{Tr}\, F. \qquad (4.3.12)$$

Then, a bundle E is μ-stable if $\mu(E') < \mu(E)$, for all coherent subsheaves E' of E.[14] In the heterotic case one has furthermore to impose $\mu(E) = 0$ and this leads to considering the semi-stability condition $\mu(E') \leq 0$.

A key ingredient for obtaining all the above results is the closure of the four-form $e^{-2\Phi} J \wedge J$, i.e. the fact that the internal non-Kähler space is balanced. Interestingly, this property is also preserved in our non-supersymmetric almost-complex backgrounds and this suggests a possible extension of the above notion of stability to our non-supersymmetric setting. This extension would then ultimately originate from the existence of the calibration structures characterizing our

---

[14] For an introduction to algebraic geometry in general and sheaves in particular see [228].

backgrounds (see e.g. [229] for a recent study of the properties of calibrated geometries). However, a proper analysis of this possibility is beyond the scope of this thesis.

We therefore end here our discussion about calibrations, and turn back to the problem of finding solutions to the conditions (4.2.24) and (4.2.26) that need to be satisfied in order to solve the EoM's of our SUSY-breaking theory.

4.4 $\frac{1}{2}$ Domain-Wall supersymmetry breaking

In this section we solve the constraint (4.2.26) by identifying, via the above dictionary relating supersymmetry conditions and calibrations, a subclass of the SUSY-breaking configurations discussed in section 4.2 with particularly interesting properties

This subclass presents a rather constrained internal geometry and SUSY-breaking pattern with respect to the general DWSB case. More precisely, via an effective four-dimensional interpretation we will show that these compactifications are particular realizations of four-dimensional no-scale vacua with broken supersymmetry. This result can be understood by translating the definition of our subclass to the type II context. Then one finds that the $\mathcal{N} = 0$ vacua described in [133] fall into this subclass.[15] We would therefore expect that, upon the usual chain of dualities, any of the $\mathcal{N} = 0$ vacua of [133] are mapped within the class of heterotic backgrounds described in this section.

4.4.1 $\frac{1}{2}$ DWSB vacua

Let us first define our subclass of backgrounds by demanding that they satisfy in addition to equations (4.2.1) the condition

$$\mathrm{Im}\left[e^{i\vartheta}\mathrm{d}\left(e^{-2\Phi}\Omega\right)\right] = 0, \qquad (4.4.1)$$

for some phase $e^{i\vartheta}$. This condition is clearly weaker than (4.1.9a), and thus trivially satisfied for a supersymmetric background. In the case where the phase $e^{i\vartheta} = e^{i\vartheta_0}$ is constant, one can think of equation (4.4.1) as half-imposing the supersymmetry equation (4.1.9a), in the sense that Ω does not satisfy (4.1.9a) but $\mathrm{Im}(e^{i\vartheta_0}\Omega)$ does. In order to separate the subclass for which (4.4.1) holds from our general ansatz we will call it $\frac{1}{2}$DWSB backgrounds. Finally, note that in terms of the differential conditions satisfied by our background, imposing equations (4.1.9b), (4.1.9c) and (4.4.1) seems as close as we may get to a supersymmetric background, since by additionally imposing (4.4.1) with the choice of $\vartheta' \neq \vartheta \bmod \pi$ the whole set of SUSY conditions (4.1.9) follows.

By decomposing (4.4.1) in terms of J and Ω it is easy to convince oneself that it is equivalent to require the condition (4.2.3), as well as $\mathrm{Im}(e^{i\vartheta}W_1) = 0$ and $\mathrm{Im}(W_2/W_1) = 0$. From the latter

---

[15]The same applies to the so-called one-parameter DWSB vacua constructed in [197].

condition one obtains that the matrix U defined in (4.2.25) is real. This in turn implies that $U^\dagger = -U$, and so the constraint (4.2.26) reads

$$U^2 = -\mathbb{1}, \quad \Leftrightarrow \quad (JU)^2 = \mathbb{1}, \tag{4.4.2}$$

where we have also used that $[U, J] = 0$. It is then natural to introduce the matrices

$$P_N = \frac{1}{2}(\mathbb{1} + IU), \qquad P_N^\perp = \frac{1}{2}(\mathbb{1} - IU), \tag{4.4.3}$$

satisfying the following properties

$$P_N^\dagger = P_N, \qquad P_N^2 = P_N, \qquad [P_N, I] = 0, \tag{4.4.4}$$

and similarly for $P_N^\perp$. These properties imply that P_N and $P_N^\perp$ are projection operators that split the tangent bundle into two orthogonal sub-bundles

$$TM = T_N \oplus T_N^\perp, \tag{4.4.5}$$

which preserve the almost complex structure in the sense that $J \cdot T_N \subset T_N$ and $J \cdot T_N^\perp \subset T_N^\perp$, and so we can write $J = J_{T_N} + J_{T_N^\perp}$, with J_{T_N} and $J_{T_N^\perp}$ almost-complex structures on T_N and $T_N^\perp$ respectively. On the other hand, (4.2.27) implies that

$$\text{Tr}\, P_N = 2, \tag{4.4.6}$$

and so T_N is a two-dimensional vector space. Finally, by using (4.2.25) one can see that

$$W_2 = 2 W_1 J \left(P_N^\perp - 2 P_N \right), \tag{4.4.7}$$

which together with (4.2.8) gives

$$e^{2\Phi} \mathrm{d}(e^{-2\Phi} \Omega) = 3 W_1 (J P_N^\perp) \wedge (J P_N^\perp). \tag{4.4.8}$$

By Frobenius' theorem[16], the subbundle T_N can be integrated into a two-dimensional submanifold. We are thus led to consider a fibered space of the form

$$N \hookrightarrow M \xrightarrow{\pi} \mathcal{B}, \tag{4.4.9}$$

---

[16]Frobenius theorem states that for an integrable subbundle of the tangentbundle TM there exists a foliation of M that is regular, i.e. the tangent bundles of the leaf equals the subbundle (see e.g. [230]).

with a two-dimensional fiber N and a four-dimensional base $\mathcal{B}$. Note that JP_N defines a preferred (integrable) complex structure j on N at each point of $\mathcal{B}$. This fibration structure induces a dual decomposition of the cotangent bundle $TM^* = T_\mathcal{B}^* \oplus T_\mathcal{B}^{*\perp}$, with $T_\mathcal{B}^* = (T_N^\perp)^*$ and $T_\mathcal{B}^{*\perp} = (T_N)^*$, and so we can decompose the SU(3) structure accordingly as

$$J = J_\mathcal{B} + j, \qquad \Omega = \Omega_\mathcal{B} \wedge \Theta. \qquad (4.4.10)$$

Here, we defined $j = JP_N = -\frac{i}{2}\Theta \wedge \overline{\Theta}$, $J_\mathcal{B} = JP_N^\perp$, etc. Using this notation, we can rewrite (4.4.8) as

$$e^{2\Phi}\mathrm{d}(e^{-2\Phi}\Omega) = 3W_1 J_\mathcal{B} \wedge J_\mathcal{B}. \qquad (4.4.11)$$

An important implication of (4.4.11) is that the fibration (4.4.9) is equipped with a transverse complex structure, i.e. an integrable complex structure along the base. To see this, let us introduce some (non-canonical) local coordinates[17] x^a, y^m, where the x^a are along the fiber N and the y^m are along the base $\mathcal{B}$, in the sense that the fibers are described by $y^m = $ const. We can then write $\Omega_\mathcal{B} = \frac{1}{2}(\Omega_\mathcal{B})_{mn}\mathrm{d}y^m \wedge \mathrm{d}y^n$ and, by using the coordinate co-frame $\mathrm{d}x^a$ and $\mathrm{d}y^m$ for TM^*, we can (non-canonically) split $\Theta = \Theta_\mathcal{B} + \Theta_N$ (with $\Theta_N \neq 0$) and $\mathrm{d} = \mathrm{d}_\mathcal{B} + \mathrm{d}_N$, with obvious notation. Now, taking components of $\mathrm{d}\Omega$ with one or two indices along the fiber, it is implied by (4.4.11) that

$$\mathrm{d}_\mathcal{B}\Omega_\mathcal{B} \wedge \Theta_N + \Omega_\mathcal{B} \wedge \mathrm{d}_\mathcal{B}\Theta_N + \Omega_\mathcal{B} \wedge \mathrm{d}_N\Theta_\mathcal{B} = 0, \quad \mathrm{d}_N\Omega_\mathcal{B} \wedge \Theta_N + \Omega_\mathcal{B} \wedge \mathrm{d}_N\Theta_N = 0. \qquad (4.4.12)$$

The first condition in (4.4.12) is telling us that $(\mathrm{d}_\mathcal{B}\Omega_\mathcal{B} - \alpha_\mathcal{B} \wedge \Omega_\mathcal{B})|_{x^a=\mathrm{const}^a} = 0$, for some one-form $\alpha_\mathcal{B} = (\alpha_\mathcal{B})_m \mathrm{d}y^m$, and this in turn means that $\Omega_\mathcal{B}$ defines an integrable complex structure on the four-dimensional slice $x^a = $ const. The second equation implies that $\partial_a \Omega_\mathcal{B} \propto \Omega_\mathcal{B}$ and thus that this complex structure is conserved while moving along the fiber coordinates x^a. Note, that even if the choice of coordinates x^a, y^m is not canonical, as one may go to a different coordinate system

$$x^a \to \tilde{x}^a(x,y), \qquad y^m \to \tilde{y}^m(y), \qquad (4.4.13)$$

the above conclusions are clearly covariant under such a change of coordinates and hold therefore in general. This means that $\Omega_\mathcal{B}$ defines an integrable complex structure on the base or, in other words, that one can introduce complex coordinates (z^1, z^2) on the base such that $\Omega_\mathcal{B} = \frac{1}{2}(\Omega_\mathcal{B})_{ij}\mathrm{d}z^i \wedge \mathrm{d}z^j$ and $\partial_a(\Omega_\mathcal{B})_{ij} \propto (\Omega_\mathcal{B})_{ij}$.

On the other hand, since N is two-dimensional the almost complex structure j of the fiber is always integrable. Our background is hence characterized by two integrable complex structures on

---

[17]Here, the indices a and m are of course two- and four-dimensional, respectively.

the base and on the fiber, respectively, which however do not combine into an integrable complex structure on the whole space. This is only possible for a non-trivial fibration. Explicit examples of such fibrations will be discussed in section 4.5.

To summarize, for $\frac{1}{2}$DWSB backgrounds the compactification manifold M can be described by a fibration of a two-dimensional fiber N over a four-dimensional base $\mathcal{B}$, both N and $\mathcal{B}$ being complex manifolds. Moreover, the torsion classes of M are given by (4.2.7) and

$$W_2 = 2W_1(J - 3j), \quad \text{with} \quad W_1 = f e^{-i\vartheta}. \quad (4.4.14)$$

Here, f is a real function and j a real (1,1)-form such that $j \cdot J = j \cdot j = 1$ (in fact, $j = JP_N$ and $J - j = J_\mathcal{B}$). The remaining torsion class W_3 is constrained via the presence of the flux H and equation (4.2.1b).

Note that in order to obtain vacua one also needs to impose the condition (4.2.24) on the two-form $\mathcal{S}$, which can be rewritten as

$$\mathcal{S} = 6W_1 J_\mathcal{B}. \quad (4.4.15)$$

The form $\mathcal{S}$ appears also in the internal gravitino variation (4.2.21b) and so the violation of the gravitino and dilatino Killing spinor equations takes the form

$$\delta\Psi_\mu = 0, \quad (4.4.16a)$$

$$\delta\Psi_m = -\frac{3i}{2} W_1 (J_\mathcal{B})_{mn} \zeta \otimes \gamma^n \eta_+^* + \text{c.c.}, \quad (4.4.16b)$$

$$\delta\lambda = \frac{3}{2} W_1 \otimes \eta_+^* + \text{c.c.}. \quad (4.4.16c)$$

As we have discussed in section 4.2.2, W_1 is directly related to the gravitino mass. $J_\mathcal{B}$ is in turn related to the source of SUSY-breaking, and more precisely to the chiral fields that develop non-vanishing F-terms, as the four-dimensional analysis of the next subsection shows.

4.4.2 Four-dimensional interpretation

In the following we would like to show that $\frac{1}{2}$DWSB vacua can be interpreted as the ten-dimensional realization of a no-scale supersymmetry breaking in four dimensions [231,232]. In order to see this, we need an effective four-dimensional supergravity describing these kinds of flux compactifications. Unfortunately, such a theory is not available. However, the problem of identifying the four-dimensional theory governing quite general heterotic flux compactifications has been investigated in several papers (see e.g. [233, 234] and references therein) which, under suitable simplifying assumptions, arrived at precise expressions for the effective four-dimensional theory. One of these assumption is that all the scalar quantities in the internal spaces are assumed to be constant. Therefore, in the following we will implicitly approximate W_1 and the dilaton to be constant, which by (4.2.7) implies that $W_4 = W_5 = 0$. Note also that W_1 constant implies that $\vartheta = \vartheta_0$ is constant, and so without loss of generality we can set it to zero. Then, our compactification manifold M satisfies $\operatorname{Im} W_1 = \operatorname{Im} W_2 = W_4 = W_5 = 0$, and thus reduce to compactifications on half-flat manifolds, like those studied in [233, 235–238].

In terms of the SU(3) structure described by J and Ω, the Kähler potential reads[18]

$$\mathcal{K} = -\log(s + \bar{s}) - \log\left(\frac{1}{3!\, l_s^6} \int_M J \wedge J \wedge J\right) - \log\left(-\frac{i}{8\, l_s^6} \int_M \Omega \wedge \overline{\Omega}\right), \qquad (4.4.17)$$

where $l_s = 2\pi\sqrt{\alpha'}$ and the real part of s is given by

$$\operatorname{Re} s = \frac{1}{2\pi l_s^6} \int \mathrm{dvol}_M\, e^{-2\Phi} = \frac{\mathrm{Vol}(M)}{2\pi l_s^6 g_s^2}\,. \qquad (4.4.18)$$

On the other hand, the superpotential [136, 137] has the standard Gukov-Vafa-Witten [141] form[19]

$$\mathcal{W} = \frac{i\, M_P^3}{8\pi\, l_s^5} \int_M \Omega \wedge (H - i\mathrm{d}J) = \frac{i\, M_P^3}{8\pi\, l_s^5} \int_M \Omega \wedge G\,. \qquad (4.4.19)$$

We can use these expressions to compute the gravitino mass from the standard four-dimensional supergravity formula

$$m_{3/2} = \frac{1}{M_P^2}\, e^{\mathcal{K}/2}\mathcal{W}\,. \qquad (4.4.20)$$

Computing $\mathcal{W}$ and $\mathcal{K}$ by using Ω and J restricted by the conditions provided in section 4.2.1,[20] and approximating the dilaton and W_1 as constants, one finds indeed agreement with (4.2.34).

In order to show the no-scale structure [231, 232] of $\frac{1}{2}$DWSB vacua we follow closely [240], where a similar analysis was done for type II theories. We expand $J = J_\mathcal{B} + j$ as

$$J_\mathcal{B} = l_s^2 (\operatorname{Re} t^a)\, \rho_a\,, \qquad j = l_s^2 (\operatorname{Re} u)\, j^{(0)}\,, \qquad (4.4.21)$$

where ρ_a is some basis of self-dual two-forms on the base $\mathcal{B}$ and $j^{(0)}$ represents a fixed reference two-form orthogonal to the base. t^a and u may be considered as pseudo-Kähler moduli of the

---

[18] Here, for simplicity, we do not consider the gauge bundle contribution to the Kähler potential.

[19] The overall factor in (4.4.19) has been fixed by reproducing (4.4.17) and (4.4.19) following the approach of [239], which combines the domain-wall arguments, analogous to the ones originally used in [141], and the use of superconformal supergravity in four dimensions.

[20] Notice that, in fact, the (3,0)-form Ω appearing in (4.4.17) and (4.4.19) has no fixed normalization and only matches the Ω used in the rest of the paper (normalized as $\Omega \wedge \overline{\Omega} = i8\mathrm{dvol}_M$) up to a overall constant. Such a change of normalization corresponds to a Kähler transformation in the four-dimensional theory and thus does not affect physical quantities like $|m_{3/2}|$, as it is clear from (4.4.17), (4.4.19), and (4.4.20).

base and the fiber respectively, complexified into chiral four-dimensional fields by the coefficients appearing in the expansion of the internal two-form B in the same basis. Analogously, the chiral field s appearing in (4.4.17) must be considered as the chiral field obtained by complexifying Re s by the axion dual to the external $B_{\mu\nu}$. By assuming off-shell the condition $d\Omega \wedge J_B = 0$, which follows from (4.4.11), it is easy to be convinced that the superpotential (4.4.19) depends only on the fiber pseudo-Kähler modulus u and pseudo-complex structure moduli z^i encoded in Ω:

$$\frac{\partial \mathcal{W}}{\partial s} = 0, \quad \frac{\partial \mathcal{W}}{\partial t^a} = 0. \tag{4.4.22}$$

On the other hand, the Kähler potential can be expanded

$$\mathcal{K} = -\log(\operatorname{Re} s) - \log[h_{ab}(\operatorname{Re} t^a)(\operatorname{Re} t^b)] - \log(\operatorname{Re} u) - \log\left(-\frac{i}{8\,l_s^6}\int_M \Omega \wedge \overline{\Omega}\right), \tag{4.4.23}$$

where

$$h_{ab} = \frac{1}{2}\int_M \rho_a \wedge \rho_b \wedge j^{(0)}. \tag{4.4.24}$$

Introducing common indices $\alpha, \beta, \ldots$ for (s, t^a), one finds

$$\mathcal{K}^{\alpha\bar{\beta}}\,\partial_\alpha \mathcal{K}\,\partial_{\bar{\beta}} \mathcal{K} = 3, \tag{4.4.25}$$

where $\mathcal{K}^{\alpha\bar{\beta}}$ is the inverse of the matrix $\partial_\alpha \partial_{\bar{\beta}}\mathcal{K}$. The conditions (4.4.22) and (4.4.25) are typical of no-scale models and indeed are sufficient to give a semi-positive definite potential

$$V_{\text{4D-SUGRA}} = e^{\mathcal{K}}\,\mathcal{K}^{I\bar{J}} D_I \mathcal{W} D_{\bar{J}} \overline{\mathcal{W}} + (\text{D-term})^2 \geq 0, \tag{4.4.26}$$

where $D_I \mathcal{W} = \partial_I \mathcal{W} + (\partial_I \mathcal{K})\mathcal{W}$, $\mathcal{K}^{I\bar{J}}$ is the inverse of $\partial_I \partial_{\bar{J}}\mathcal{K}$, and $I, J, \ldots$ are indices collectively denoting the chiral fields (u, z^i).

In order to extremize the potential, one needs to impose

$$D_u \mathcal{W} = 0, \quad D_i \mathcal{W} = 0, \quad (\text{D-term}) = 0. \tag{4.4.27}$$

Explicitly, we have the expressions

$$D_u \mathcal{W} \propto \frac{1}{u + \bar{u}}\int_M \Omega \wedge \overline{G}, \tag{4.4.28a}$$

$$D_i \mathcal{W} \propto \int_M \xi_i \wedge G, \tag{4.4.28b}$$

where
$$\xi_i = \partial_i \Omega - \Omega \frac{\int_M \partial_i \Omega \wedge \overline{\Omega}}{\int_M \Omega \wedge \overline{\Omega}} \quad (4.4.29)$$

should be the basis of $(2,1)$-forms relevant for the four-dimensional description. Then, assuming that the truncated theory makes sense, imposing (4.4.27) is equivalent to the conditions (4.2.10), which were obtained from our previous ten-dimensional analysis. We can then interpret (4.2.10) as the above four-dimensional F-flatness conditions. In addition the F-terms that do not enter the scalar potential, $D_s \mathcal{W}$ and $D_a \mathcal{W}$, are non-vanishing whenever $G^{0,3} \neq 0$ (or equivalently $W_1 \neq 0$), again in agreement with the ten-dimensional analysis. The only remaining ten-dimensional condition is (4.2.1a), which becomes $J \wedge dJ = 0$ in the constant dilaton approximation, and which can be interpreted as the four-dimensional D-flatness condition.

Finally, let us also briefly consider the gauge bundle sector. By a simple dimensional reduction of the ten-dimensional action (2.2.1), it is easy to see that the kinetic term for the four-dimensional gauge field is given by

$$-\frac{1}{4} \operatorname{Re} s \operatorname{Tr} F^{\mu\nu} F_{\mu\nu} \quad (4.4.30)$$

Therefore, the holomorphic gauge coupling is given by $f(s) = s$, and from the standard formula for the gaugino mass we get

$$m_{1/2}^{4D} = -\frac{1}{M_P^2} e^{\mathcal{K}/2} \mathcal{K}^{s\bar{s}} D_s \mathcal{W} \, \partial_{\bar{s}} \log(\operatorname{Re} f) = \frac{1}{M_P^2} e^{\mathcal{K}/2} \mathcal{W} = m_{3/2} \,. \quad (4.4.31)$$

This agrees with (4.2.38), which was obtained directly by dimensionally reducing the fermionic ten-dimensional action, and shows that our $\frac{1}{2}$DWSB ansatz is consistent with the more general picture developed in section 4.2.

After having found the four-dimensional interpretation of the $\frac{1}{2}$DWSB ansatz, it is still necessary to find explicit solutions. This will be the topic of the next section.

4.5 Examples via homogeneous fibrations

In order to illustrate the general features of $\frac{1}{2}$DWSB vacua, we will discuss a concrete setting in this section in which explicit examples can be constructed. Recall that the $\frac{1}{2}$DWSB ansatz implies that the compactification manifold M is based on a fibration of the form

$$N \hookrightarrow M \xrightarrow{\pi} \mathcal{B}, \quad (4.5.1)$$

with a two-dimensional fiber N and a four-dimensional base $\mathcal{B}$. We will at first simplify the geometry by assuming that all geometric quantities are only base-dependent. In other words, we assume that Φ, W_1, and the forms $\Omega_\mathcal{B}$ and $J_\mathcal{B}$ in (4.4.10) can be seen as functions and forms on the base $\mathcal{B}$.[21] This implies that $\mathrm{d}\Phi$, $\mathrm{d}W_1$, $\mathrm{d}\Omega_\mathcal{B}$, and $\mathrm{d}J_\mathcal{B}$ are also solely forms on the base $\mathcal{B}$.

This simplifying assumption has several consequences. For instance, by (4.4.11) one can see that the pull-back of $\mathrm{d}\Theta$ to any fiber N vanishes, $\mathrm{d}\Theta|_N = 0$. This means that the pulled-back hermitian metric $g|_N$ on N is flat, and so we are led to take a two-torus as a fiber

$$N \simeq T^2, \tag{4.5.2}$$

i.e. M is elliptically fibered. Starting with [143], such elliptically fibered manifolds have played a key role in the literature of torsional heterotic backgrounds, and in particular in constructions motivated by duality arguments – see e.g. [189, 190, 241, 242]. Here, we see the elliptic fibration arising from imposing a rather simple pattern of torsional supersymmetry breaking. In the following we will analyze which further constraints this pattern imposes on M.

4.5.1 Constraints on the elliptic fibration

Since our fiber is a two-torus, the one-form Θ, introduced in (4.4.10), takes the form

$$\Theta = \frac{l_s L_{T^2}}{\sqrt{\mathrm{Im}\,\tau}} e^C \theta, \qquad \theta = \eta^1 - \tau \eta^2, \tag{4.5.3}$$

where $l_s = 2\pi\sqrt{\alpha'}$ and η^a ($a = 1, 2$) are one-forms which can be written locally as

$$\eta^a = \mathrm{d}x^a + A^a(y), \tag{4.5.4}$$

with $x^a \simeq x^a + 1$ dimensionless coordinates along the T^2-fiber and $A^a(y)$ one-forms along $\mathcal{B}$, that only depend on the base coordinates y^α, $\alpha = 1, \ldots, 4$. τ is the complex structure of the T^2 fiber, L_{T^2} is the dimensionless T^2 length scale in string units, and $\langle e^C \rangle \lesssim 1$ encodes the non-trivial dependence of the fiber volume on the base coordinates, which is given by $\mathrm{Vol}(T^2) = e^{2C} l_s^2 L_{T^2}^2$.

The one-forms $A^a(y)$ can be seen locally as U(1) gauge fields on $\mathcal{B}$, while globally they can be further twisted by SL(2, $\mathbb{Z}$) transformations, the large diffeomorphism group of T^2 if the T^2-fibration degenerates at some points. The same applies to the associated U(1) field strengths $\omega^a = \mathrm{d}A^a$, which must obey an SL(2, $\mathbb{Z}$)-twisted quantization condition and so define SL(2, $\mathbb{Z}$)-twisted cohomology classes in $\mathcal{B}$. Note however that ω^a are cohomologically trivial in the ambient manifold M, since $\mathrm{d}\eta^a = \omega^a$.

Let us next determine how the background quantities J and Ω, decomposed as in (4.4.10), are

---
[21]Stated more precisely, we assume that Φ, W_1, $\Omega_\mathcal{B}$, $J_\mathcal{B}$ and F can be obtained as the pull-back of corresponding functions and forms by the fibration map $\pi : M \to \mathcal{B}$.

constrained by our $\frac{1}{2}$DWSB ansatz.[22] Using the fact that the right hand side of (4.4.11) has only legs along the base, one arrives at the conditions

$$\mathrm{d}\left(\frac{e^{C-2\Phi}}{\sqrt{\mathrm{Im}\,\tau}}\Omega_{\mathcal{B}}\right) = 0, \qquad (4.5.5\mathrm{a})$$

$$\bar{\partial}\tau = 0, \qquad (4.5.5\mathrm{b})$$

while equation (4.2.1a) yields

$$\mathrm{d}(e^{2C-2\Phi}J_{\mathcal{B}}) = 0, \qquad (4.5.5\mathrm{c})$$

$$J_{\mathcal{B}} \wedge \chi = 0. \qquad (4.5.5\mathrm{d})$$

Here, we have introduced the complex two-form

$$\chi = \omega^1 - \tau\omega^2 = \mathrm{d}\theta + \mathrm{d}\tau \wedge \eta^2. \qquad (4.5.6)$$

From (4.5.5a) and (4.5.5c) we see that $\mathcal{B}$ not only admits an integrable complex structure, but also a Kähler structure, with holomorphic (2,0)-form and Kähler form given by

$$\hat{\Omega}_{\mathcal{B}} = \frac{e^{2D-C}}{\sqrt{\mathrm{Im}\,\tau}}\Omega_{\mathcal{B}}, \qquad \hat{J}_{\mathcal{B}} = e^{2D}J_{\mathcal{B}}, \qquad (4.5.7)$$

and where

$$e^D = g_s\, e^{C-\Phi}, \qquad (4.5.8)$$

with g_s defined in (2.4.35). It is then natural to express the internal metric in terms of this four-dimensional Kähler metric $\mathrm{d}\hat{s}_{\mathcal{B}}^2$

$$\mathrm{d}s_M^2 = e^{-2D}\mathrm{d}\hat{s}_{\mathcal{B}}^2 + l_s^2 L_{T^2}^2 \frac{e^{2C}}{\mathrm{Im}\,\tau}\theta\otimes\bar{\theta}. \qquad (4.5.9)$$

However, the metric $\mathrm{d}\hat{s}_{\mathcal{B}}^2$ is Calabi-Yau only if $e^{2C}\mathrm{Im}\,\tau$ is constant, as

$$\hat{J}_{\mathcal{B}} \wedge \hat{J}_{\mathcal{B}} = \frac{1}{2}e^{2C}\mathrm{Im}\,\tau\,\hat{\Omega}_{\mathcal{B}} \wedge \overline{\hat{\Omega}}_{\mathcal{B}}. \qquad (4.5.10)$$

---

[22]To make our conventions compatible with those usually adopted in the literature, we take the choices

$$J_{\mathcal{B}} = -e^1 \wedge e^2 - e^3 \wedge e^4, \qquad j = e^5 \wedge e^6, \qquad \Omega_{\mathcal{B}} = (e^1+ie^2)\wedge(e^3+ie^4), \qquad \Theta = e^5 - ie^6,$$

where $e^1,\ldots,e^6$ is an oriented orthonormal coframe on M. Indeed, note that with these choices $J_{\mathcal{B}}$ and Ω_B are self-dual under Hodge duality on the base: $*_{\mathcal{B}}J_{\mathcal{B}} = J_{\mathcal{B}}$, $*_{\mathcal{B}}\Omega_{\mathcal{B}} = \Omega_{\mathcal{B}}$.

Imposing the condition (4.2.1b) leads to the following expression for the three-form H

$$H = \hat{*}_B \mathrm{d}\, e^{-2D} - l_s^2\, L_{T^2}^2 \left[(\mathrm{d}^c e^{2C}) \wedge \eta^1 \wedge \eta^2 + \frac{e^{2C}}{\mathrm{Im}\,\tau} \mathrm{Re}\big(\hat{*}_B \chi \wedge \overline{\theta}\big) \right], \qquad (4.5.11)$$

where $\mathrm{d}^c = i(\partial - \overline{\partial})$ and $\hat{*}_B$ is the Hodge star belonging to the Kähler metric $\mathrm{d}\hat{s}_B^2$. This three-form flux should satisfy the BI (4.3.1) and appropriate quantization conditions, respectively, in order to give consistent vacua. We will discuss these issues below.

Note that the above constraints would also apply to any supersymmetric background based on an homogeneous elliptic fibration, due to the fact that we considered only the parts of (4.4.11) that have components along the fiber. Indeed, the fact that our background breaks supersymmetry is purely encoded in the condition

$$\mathrm{d}(e^{-2\Phi}\Omega) = g_s^{-2}\, l_s\, L_{T^2}\, \hat{\Omega}_B \wedge \chi \neq 0, \qquad (4.5.12)$$

which arises from the purely base dependent part of (4.4.11). In order to break supersymmetry in this way, one must require that χ has non-vanishing (0,2) and (2,0) components.[23] Of course, by relaxing the $\tfrac{1}{2}$DWSB ansatz further ways of breaking supersymmetry arise. For instance, we see from (4.5.5d) that χ must be primitive and from section 4.4.2 that a non-primitive χ can be interpreted as a non-vanishing D-term.

Recall now that for this class of $\mathcal{N} = 0$ vacua one needs to impose the residual condition (4.2.24) coming from the equation of motions. For our elliptic fibration, this is equivalent to require that $\partial(e^{-2D} W_1) = 0$, which is solved by

$$W_1 = c_{\mathrm{SB}}\, e^{2D}, \qquad (4.5.13)$$

where c_{SB} is a constant, which parametrizes the amount of SUSY-breaking and is therefore proportional to the gravitino mass.

Gravitino mass

The parameter c_{SB} should directly enter physical quantities, which measure the amount of SUSY-breaking of a compactification, as it is the case for the gravitino mass. Following our discussion of section 4.2.2 we will first consider the gravitino mass density $\mathfrak{m}_{3/2}$. By comparing (4.5.13) and (4.2.32) it reads

$$\mathfrak{m}_{3/2} = 3\, e^A\, c_{\mathrm{SB}}\, e^{2D}. \qquad (4.5.14)$$

---

[23]Having $\chi^{0,2} \neq 0$, but $\chi^{2,0} = 0$ is not sufficient, since we can change the orientation of T^2, basically swapping χ and $\overline{\chi}$, and thus getting an $\mathcal{N} = 1$ supersymmetry. Having $\chi^{0,2} = \chi^{2,0} = 0$ means that both orientations on T^2 lead to preserved supersymmetry, and thus we have an $\mathcal{N} = 2$ compactification.

Since e^A is constant, we see that the SUSY-breaking is milder in the points of $\mathcal{B}$ with strong conformal factor $e^{-2D} \gg 1$. A rough estimate of the gravitino mass is obtained by the approximation $e^C \simeq 1$ and e^D, τ constant. Then, by using (4.4.11), (4.5.12), and (4.5.13) we find

$$c_{\text{SB}} \simeq \frac{2l_s\, L_{T^2}}{3 \operatorname{Im} \tau} \times \frac{\int_{\mathcal{B}} \hat{\Omega}_{\mathcal{B}} \wedge \chi}{\int_{\mathcal{B}} \hat{\Omega}_{\mathcal{B}} \wedge \overline{\hat{\Omega}_{\mathcal{B}}}}\,. \tag{4.5.15}$$

Taking into account (2.4.34), (4.2.33), and (4.5.14), one finally gets

$$m_{3/2} \simeq \frac{g_s\, M_{\text{P}}\, e^{4D}\, \|\chi^{0,2}\|}{2\, L_{\mathcal{B}}^4\, \sqrt{\pi \operatorname{Im} \tau}}\,, \tag{4.5.16}$$

where

$$L_{\mathcal{B}}^4 \;=\; l_s^{-4}\, \widehat{\operatorname{Vol}}(\mathcal{B})\,, \tag{4.5.17}$$

and where we have introduced the quantity

$$\|\chi^{0,2}\| \;=\; \frac{\int_{\mathcal{B}} \hat{\Omega}_{\mathcal{B}} \wedge \chi}{\sqrt{\left(\int_{\mathcal{B}} \hat{\Omega}_{\mathcal{B}} \wedge \overline{\hat{\Omega}_{\mathcal{B}}}\right)}}\,, \tag{4.5.18}$$

which measures the alignment of χ with $\overline{\hat{\Omega}_{\mathcal{B}}}$. While a generalization of the above expression for non-trivial e^Φ, e^C, e^D, and τ is straightforward, (4.5.16) already captures most of the qualitative behavior of $m_{3/2}$, and is hence sufficient for the purposes of our discussion.

From (4.5.16) we see that one can suppress the SUSY-breaking scale by combining the following (possibly non-independent) conditions: $g_s \ll 1$, $\|\chi^{0,2}\| \ll 1$, $L_{\mathcal{B}}^4 \gg 1$, $e^D \ll 1$ and $\operatorname{Im} \tau \gg 1$. Recall that we are implicitly assuming that $\chi^{2,0} \neq 0$, which already selects an $\mathcal{N} = 1$ supersymmetry in four-dimensions, eventually broken by the non-vanishing $\chi^{0,2}$. The expression (4.5.16) then refers to the gravitino which is selected by the flux $\chi^{2,0}$.

Bianchi identity and tadpole conditions

We would like to impose the Bianchi Identity (4.3.1) now. From (4.5.11) we obtain

$$\begin{aligned}
\mathrm{d}H &= \mathrm{d}\hat{*}_{\mathcal{B}}\, \mathrm{d}\, e^{-2D} + l_s^2\, L_{T^2}^2 \bigg\{ (\mathrm{d}^c \mathrm{d}\, e^{2C}) \wedge \eta^1 \wedge \eta^2 - \frac{e^{2C}}{\operatorname{Im} \tau}\, \hat{*}_{\mathcal{B}}\, \chi \wedge \overline{\chi} \\
&\quad - \operatorname{Re}\!\left[\mathrm{d}\!\left(\frac{e^{2C}}{\operatorname{Im} \tau}\, \hat{*}_{\mathcal{B}} \chi\right) \wedge \overline{\theta} - \frac{e^{2C}}{\operatorname{Im} \tau}\, \hat{*}_{\mathcal{B}} \chi \wedge (\overline{\partial \tau}) \wedge \eta^2 \right] + \frac{\mathrm{d}^c e^{2C} \wedge \operatorname{Im}(\chi \wedge \overline{\theta})}{\operatorname{Im} \tau} \bigg\} \\
&= \frac{\alpha'}{4}\, (\operatorname{Tr} R_+ \wedge R_+ - \operatorname{Tr} F \wedge F) + 2\kappa^2\, \tau_{\text{NS5}}\, \delta^4(\Pi_2)\,.
\end{aligned} \tag{4.5.19}$$

Note that dH has components that are totally restricted to the base as well as contributions with one or even two legs along the fiber. This means that the sources appearing in the last line of (4.5.19) also need such contributions. For example, in order to cancel the contribution of $(d^c d\, e^{2C}) \wedge \eta^1 \wedge \eta^2$ one needs sources along the T^2-fiber. Such sources could be gauge bundles with $\text{Tr}(F \wedge F)$ dual to a two-cycle in $\mathcal{B}$, or NS5-branes that do not wrap the T^2-fiber. For instance, an NS5-brane wrapping besides the four-dimensional spacetime a holomorphic curve $\Pi_2 \subset \mathcal{B}$ would contribute a factor of $\delta^2_{\mathcal{B}}(\Pi_2) \wedge \eta^1 \wedge \eta^2$ to the right hand side of (4.3.1), leading to $d^c d\, e^{2C} = \delta^2_{\mathcal{B}}(\Pi_2)$. On the other hand, configurations in which sources are located along one cycle of T^2 and along a one-dimensional cycle of $\mathcal{B}$ could cancel the middle line of (4.5.19).

Since all this would be generically difficult to treat, we make the simplifying assumption that only sources completely located on the base should contribute to the Bianchi identity. This means that we will only consider NS5-branes that wrap the T^2-fiber and are sitting at points p_i of the base. We furthermore assume that the T^2 fibration has constant volume, so that e^C is constant along $\mathcal{B}$. In particular we set

$$e^C = 1, \quad d\left(\frac{\hat{*}_{\mathcal{B}}\text{Re}\,\chi}{\text{Im}\,\tau}\right) = 0, \quad d\left(\frac{\hat{*}_{\mathcal{B}}\text{Re}(\bar{\tau}\chi)}{\text{Im}\,\tau}\right) = 0 \ . \quad (4.5.20)$$

We can further motivate these restrictions by considering the $\text{Tr}\, R_+ \wedge R_+$-term in (4.5.19). Following [242] and expanding the curvature tensor R_+ in powers of the base length scale $L_{\mathcal{B}}$ and the fiber length scale L_{T^2} we obtain

$$\text{Tr}\, R_+ \wedge R_+ = \text{Tr}\, R_{\mathcal{B}+} \wedge R_{\mathcal{B}+} + \mathcal{O}\left(\frac{L^2_{T^2}}{L^2_{\mathcal{B}}}\right) \ . \quad (4.5.21)$$

Here $R_{\mathcal{B}+}$ is the torsionful curvature of the base $\mathcal{B}$ computed using the four-dimensional metric $e^{-2D} d\hat{s}^2_{\mathcal{B}}$. This means that $\text{Tr}\, R_+ \wedge R_+$ is a four-form on $\mathcal{B}$ as long as the ratio of fiber size to base size is small.[24]

But by (4.5.16) we can suppress the SUSY-breaking scale by considering an anisotropic fibration, with the base much larger then the fiber

$$L_{\mathcal{B}} \gg L_{T^2} \ . \quad (4.5.22)$$

Moreover, by the discussion of section 4.2.3, an anisotropic fibration should simplify the conditions on the curvature (4.2.29) as discussed below. We see therefore that the assumption that (4.5.19) should only live on the base is justified if one wants to have a mild SUSY-breaking.

---

[24]Note that R_+ is invariant under an overall rescaling of the six-dimensional metric and is constructed from the torsionful connection $\Gamma^i_{+jk} = \Gamma^i_{jk} + \frac{1}{2} H^i{}_{jp}$, which depends quadratically on $L_{\mathcal{B}}$ and L_{T^2}. Hence, it only depends on even powers of $L_{\mathcal{B}}/L_{T^2}$.

Then, one should also consider the gauge bundle F on M to be the pull-back of a gauge bundle $F_\mathcal{B}$ on $\mathcal{B}$. The pseudo-HYM equation (4.2.28) reduces to a anti-self duality condition on the base

$$*_\mathcal{B} F_\mathcal{B} = -F_\mathcal{B}, \qquad (4.5.23)$$

which is equivalent to the statement that $F_\mathcal{B}$ is $(1,1)$ and primitive.

Taking all this into account one can derive the following equation from (4.5.19)

$$\mathrm{d}\hat{*}_{(0)} \mathrm{d}\, e^{-2D} = \frac{1}{16\pi^2 L_\mathcal{B}^2}\big[\mathrm{Tr}(\hat{R}_\mathcal{B}\wedge \hat{R}_\mathcal{B}) - \mathrm{Tr}(F_\mathcal{B}\wedge F_\mathcal{B})\big] + \frac{1}{L_\mathcal{B}^2}\sum_i \delta^4_\mathcal{B}(p_i) \\ + \frac{1}{16\pi^2 L_\mathcal{B}^2}\mathrm{d}\,\Delta + \frac{L_{T^2}^2}{L_\mathcal{B}^2 \mathrm{Im}\,\tau}\hat{*}_{(0)}\chi\wedge\overline{\chi} + \mathcal{O}\left(\frac{L_{T^2}^2}{L_\mathcal{B}^4}\right), \qquad (4.5.24)$$

where

$$\Delta = 2\hat{*}_{(0)}\mathrm{d}\,\hat{\nabla}_{(0)}^2 D + \hat{*}_{(0)}\Big[2(\hat{\nabla}_{(0)}^2 e^{-2D})\mathrm{d}\, e^{2D} + \mathrm{d}\big(e^{2D}\hat{\nabla}_{(0)}^2 e^{-2D}\big)\Big]. \qquad (4.5.25)$$

In order to arrive at this equation one has to take two steps. One first has to reexpress R_+ in terms of the the Levi-Civita curvature on the base $R_\mathcal{B}$

$$\mathrm{Tr}\, R_{\mathcal{B}+}\wedge R_{\mathcal{B}+} = \mathrm{Tr}\,\hat{R}_\mathcal{B}\wedge\hat{R}_\mathcal{B} + 2\,\mathrm{d}\,\hat{*}_\mathcal{B}\mathrm{d}\hat{\nabla}^2 D + \mathrm{d}\,\hat{*}_\mathcal{B}\Big[2(\hat{\nabla}^2 e^{-2D})\mathrm{d}\, e^{2D} + \mathrm{d}\big(e^{2A}\hat{\nabla}^2 e^{-2D}\big)\Big]. \qquad (4.5.26)$$

Then, one introduces a dimensionless and $\mathcal{O}(\alpha'^0)$ metric $\mathrm{d}\hat{s}^2_{(0)}$ by

$$\mathrm{d}\hat{s}^2_\mathcal{B} = l_s^2 L_\mathcal{B}^2 \, \mathrm{d}\hat{s}^2_{(0)}, \qquad (4.5.27)$$

in order to make the dependence on $L_\mathcal{B}$ clear. One should also note that all 'hatted' quantities are computed using the Kähler metric $\mathrm{d}\hat{s}^2_\mathcal{B}$.

The reformulated Bianchi identity (4.5.24) will clearly only then admit a solution (up to higher order corrections in $1/L_\mathcal{B}$) if its integrated counterpart

$$N_{\mathrm{NS5}} + Q_{\mathrm{NS5}}(E) + L_{T^2}^2 \int \frac{*_\mathcal{B}\chi\wedge\overline{\chi}}{\mathrm{Im}\,\tau} = Q_{\mathrm{NS5}}(\mathcal{B}) \qquad (4.5.28)$$

is also satisfied. Here, N_{NS5} is the number of NS5-branes and we introduced the total NS5-brane charge sourced by the gauge bundle and the curvature of the base

$$Q_{\mathrm{NS5}}(E) = -\frac{1}{16\pi^2}\int \mathrm{Tr}(F_\mathcal{B}\wedge F_\mathcal{B}), \quad Q_{\mathrm{NS5}}(\mathcal{B}) = -\frac{1}{16\pi^2}\int \mathrm{Tr}(\hat{R}_\mathcal{B}\wedge\hat{R}_\mathcal{B}). \qquad (4.5.29)$$

The absence of anti-NS5-branes implies that N_{NS5} is always positive, and from (4.5.23) the same applies to $Q_{\mathrm{NS5}}(E)$. The left hand side of (4.5.28) is then always positive, and this implies an upper bound for the number of NS5-branes and non-trivial gauge bundle that can be introduced for a fixed manifold $\mathcal{B}$.

Once the condition (4.5.28) is satisfied, one can integrate (4.5.24) perturbatively. More precisely, along the lines of the $\mathcal{N} = 2$ case discussed in [190], one can rewrite (4.5.24) in terms of a shifted conformal factor

$$e^{-2D'} = e^{-2D} - \frac{1}{8\pi^2 L_\mathcal{B}^2} \hat{\nabla}_{(0)}^2 D . \qquad (4.5.30)$$

In this way (4.5.24) takes the form of a standard Poisson equation

$$\begin{aligned}-\hat{\nabla}_{(0)}^2 e^{-2D'} \hat{*}_{(0)} 1 &= \frac{1}{16\pi^2 L_\mathcal{B}^2} \left[\text{Tr}(\hat{R}_\mathcal{B} \wedge \hat{R}_\mathcal{B}) - \text{Tr}(F_\mathcal{B} \wedge F_\mathcal{B}) \right] + \frac{1}{L_\mathcal{B}^2} \sum_i \delta_\mathcal{B}^4(p_i) \\ &+ \frac{L_{T^2}^2}{L_\mathcal{B}^2 \, \text{Im}\,\tau} \hat{*}_{(0)} \chi \wedge \overline{\chi} + \mathcal{O}\left(\frac{L_{T^2}^2}{L_\mathcal{B}^4}\right) , \end{aligned} \qquad (4.5.31)$$

where on the right hand side of (4.5.31) we have omitted terms like

$$\frac{1}{16\pi^2 L_\mathcal{B}^2} \mathrm{d}\, \hat{*}_{(0)} \left[2\, (\hat{\nabla}_{(0)}^2 e^{-2D'}) \mathrm{d}\, e^{2D'} + \mathrm{d}\left(e^{2D'} \hat{\nabla}_{(0)}^2 e^{-2D'}\right) \right] . \qquad (4.5.32)$$

If (4.5.28) is fulfilled, (4.5.31) can always be integrated. As one can see by inserting (4.5.31) the possible corrections provided by (4.5.32) are then of order $\mathcal{O}(L_{T^2}^2/L_\mathcal{B}^4)$, and one can therefore consistently neglect them.[25] This means that also in the $\mathcal{N} = 0$ case there is a mechanism to stabilize the dilaton, like the one discussed in [190] for $\mathcal{N} = 2$ vacua.

The K3 case and H-flux quantization

We will now even further restrict our setting by demanding that also τ is constant. Recall that (4.5.10) implies that in this case the base should be a Calabi-Yau two-fold: $\mathcal{B} =$ K3. Furthermore, since τ does not degenerate, the one-forms $A^a(y)$ can be seen as proper U(1) gauge fields along K3 and then the corresponding field-strengths $\omega^a = \mathrm{d} A^a$ are quantized as

$$\int_{\Pi_2 \subset \mathrm{K3}} \omega^a \in \mathbb{Z} , \qquad (4.5.33)$$

and so the forms ω^a define non-trivial elements of the integral cohomology group $H^2(\mathrm{K3}, \mathbb{Z})$. In fact, from (4.5.20) we have that $\mathrm{d}(\hat{*}_{\mathrm{K3}} \chi) = 0$, and so χ must be harmonic. Finally, in order to

---

[25] Here we are implicitly ignoring the fact that, in the vicinity of NS5-branes, $\mathrm{d}\hat{*}_{(0)} \mathrm{d}\, e^{-2D}$ diverges and the tree-level supergravity approximation breaks down. However, the SUSY-breaking effects are at the $L_\mathcal{B}$ scale and very close to the NS5-brane supersymmetry is restored. Thus, we expect that NS5-brane sources can be consistently incorporated.

evaluate (4.5.28) one has to use $Q_{\text{NS5}}(\text{K3}) = -\frac{1}{2}p_1(\text{K3}) = 24$.[26]

We are thus led to the setting of non-degenerate T^2 fibrations over K3

$$T^2 \hookrightarrow M \to \text{K3}, \qquad (4.5.34)$$

which is often considered in the construction of heterotic torsional backgrounds [143, 189, 190, 241, 242]. Note in particular that for this case the SUSY-breaking conditions discussed below (4.5.12) reduce to those identified in [190] by direct inspection of the Killing spinor equations and of the $\mathcal{O}(\alpha'^0)$ equations of motion.

The K3 example allows us to discuss the quantization of the H-flux in a rather simple way. In general this is a complicated problem, partly because of the non-closure of H due to the contributions on the right hand side of (4.3.1). However, in the simplified setting of an elliptically fibered K3, the H-field (4.5.11) reduces to

$$H = \hat{*}_{\text{K3}} \mathrm{d} \, e^{-2D} - \frac{l_s^2 L_{T^2}^2}{\operatorname{Im}\tau} \operatorname{Re}(\hat{*}_{\text{K3}} \chi \wedge \overline{\theta}) \,. \qquad (4.5.35)$$

The flux H can then be written as $H = \pi^*(H_{\text{K3}}) + \pi^*(h_a) \wedge \eta^a$. H_{K3} and h_a define forms on K3 that can be pulled-back to M via π^*, the operator induced by the projector $\pi : M \to \text{K3}$. In particular we have that

$$h_1 = -\frac{l_s^2 L_{T^2}^2}{\operatorname{Im}\tau} \operatorname{Re}(\hat{*}_{\text{K3}} \chi) \quad \text{and} \quad h_2 = \frac{l_s^2 L_{T^2}^2}{\operatorname{Im}\tau} \operatorname{Re}(\overline{\tau} \hat{*}_{\text{K3}} \chi) \qquad (4.5.36)$$

are also harmonic forms. The proper quantization condition is to impose that both two-forms $l_s^{-2} h_a$ must be harmonic representatives of integral cohomology classes in $H^2(\text{K3}; \mathbb{Z})$.[27] More precisely, we get the following condition on χ

$$\frac{L_{T^2}^2}{\operatorname{Im}\tau} \int_{\Pi_2} \operatorname{Re}(\hat{*}_{\text{K3}} \chi) \in \mathbb{Z}, \qquad \frac{L_{T^2}^2}{\operatorname{Im}\tau} \int_{\Pi_2} \operatorname{Re}(\overline{\tau} \hat{*}_{\text{K3}} \chi) \in \mathbb{Z}, \qquad (4.5.37)$$

for any two-cycle $\Pi_2 \subset \text{K3}$.

Curvature corrections

Let us finally consider the R_+-dependent terms in the scalar potential (4.1.6b) for this simplified case, in order to illustrate our general discussion of section 4.2.3. First of all, by using (4.5.15) and

---

[26] Note that in our conventions $p_1(\mathcal{B}) = \frac{1}{8\pi^2} \int \operatorname{Tr}(\hat{R}_{\mathcal{B}} \wedge \hat{R}_{\mathcal{B}})$, which apparently differs by an overall sign from the standard definition of Pontjagin classes, since we use a positive-definite trace $\operatorname{Tr} = -\operatorname{Tr}_{\text{standard}}$.

[27] The h_a could be seen as U(1) field-strengths in the eight-dimensional theory obtained by compactifying the ten-dimensional theory on T^2, and are thus appropriately quantized.

taking into account that $\hat{\Omega}_{K3}$ scales as $L_{K3}^2/\sqrt{\mathrm{Im}\,\tau}$, we see that c_{SB} scales as L_{T^2}/L_{K3}^2 (assuming an approximately square T^2-fiber). Furthermore, since the fiber is flat, the leading contribution to the curvature has its origin in K3. The dimensionless length scale L_{KK} introduced in section 4.2.3 can then be identified with $e^{-D}L_{K3}$. By comparing the two estimates for the torsion class W_1, (4.2.43) and (4.5.13), we see that $L_{SB} \simeq L_{KK}^2$. Hence, we find $\beta = 1/2$ in (4.2.48).

The explicit calculation shows that the curvature terms in (4.2.50) lead indeed to a behavior as in (4.2.51)

$$|J_{ij} R_-^{ij}|^2 \sim e^{8D} |c_{SB}|^4 \,, \tag{4.5.38a}$$

$$\begin{aligned}|\Omega_{ijk} R_-^{ij}|^2 &\sim e^{6D} |c_{SB}|^2 |dD|_{K3}^2 + e^{8D} |c_{SB}|^2 \\ &\sim e^{6D} L_{K3}^{-6} + e^{8D} L_{K3}^{-8}\,. \end{aligned} \tag{4.5.38b}$$

Then, by (4.2.52) we see that the $\mathcal{O}(\alpha')$ correction to the equations of motion goes like

$$\begin{aligned}(\mathrm{EoM})_{\mathcal{O}(\alpha')} &\sim e^{4D} |c_{SB}|^2 L_{K3}^2 \big(|dD|_{K3}^2 + e^{2D} |c_{SB}|^2\big) \\ &\sim e^{4D} L_{K3}^{-4} + e^{6D} L_{K3}^{-6}\,, \end{aligned} \tag{4.5.39}$$

where in the last step, we have used the above estimate $c_{SB} \sim L_{K3}^{-2}$ and the fact that from (4.5.24) we can assume that $|dD|_{K3}^2 \sim L_{K3}^{-2}$. This already confirms the estimate made in section 4.2.3 that the contribution of the first-order potential goes like L_{KK}^{-4}, and is thus $\mathcal{O}(\alpha'^2)$. Actually, one could have D approximately constant, e.g. $|dD|^2 \lesssim L_{K3}^{-4}$, in most of the internal space. Then, the correction would be even of $\mathcal{O}(\alpha'^3)$ and so one can safely neglect the contributions from the R_+-terms to the scalar potential for our $\frac{1}{2}$DWSB models.

Putting all these pieces together, we will show in the next subsection how one can stabilize the fiber size by considering very simple examples.

4.5.2 Simple examples

In order to obtain simple explicit $\mathcal{N} = 0$ examples of the setting provided by the T^2 fibration over K3 reviewed above, let us set $N_{NS5} = 0$ and take a trivial gauge bundle on K3. This means that the condition (4.5.28) reduces to

$$L_{T^2}^2 \int \frac{\hat{*}_B \chi \wedge \overline{\chi}}{\mathrm{Im}\,\tau} = 24\,. \tag{4.5.40}$$

The task is therefore to find a primitive harmonic form χ which has both $(2,0)$ and $(0,2)$ non-vanishing components that satisfies the above equation.

The well known properties of K3 – see e.g. [243] for a review – greatly help in this search.

$H^2(K3; \mathbb{R})$ has dimension $b_2 = 22$ and, picking up a basis $\{e_I\}_{I=1}^{22}$, the inner product matrix

$$\mathcal{I}_{IJ} = \int_{K3} e_I \wedge e_J \tag{4.5.41}$$

has signature $(3, 19)$. In particular, one can choose an integral basis $\{\alpha_I\}_{I=1}^{22}$ of $H^2(K3; \mathbb{Z})$ such that

$$\mathcal{I}_{IJ} = \begin{pmatrix} 0 & 1 \\ 1 & 0 \end{pmatrix} \oplus \begin{pmatrix} 0 & 1 \\ 1 & 0 \end{pmatrix} \oplus \begin{pmatrix} 0 & 1 \\ 1 & 0 \end{pmatrix} \oplus (-E_8) \oplus (-E_8), \tag{4.5.42}$$

where E_8 is the Cartan matrix of the E_8 algebra. $\mathrm{Re}\,\hat{\Omega}_{K3}$, $\mathrm{Im}\,\hat{\Omega}_{K3}$, and $\hat{J}_{K3}$ provide a basis of the self-dual harmonic forms in $H^2(K3; \mathbb{R})$, which is a space-like plane with respect to the metric (4.5.41).

Instead of attempting a detailed general discussion of the constraints derived above, we will just provide a couple of simple examples, which should nevertheless give an idea of the qualitative features of more general solutions. First, let us take a simple choice for $\hat{\Omega}_{K3}$:

$$\hat{\Omega}_{K3} = \frac{(2\pi)^2 \alpha' L_{K3}^2}{\sqrt{\mathrm{Im}\,\tau}} (\alpha_1 + i\alpha_2), \tag{4.5.43}$$

with $\alpha_1 = e_1 + e_2$ and $\alpha_2 = e_3 + e_4$, in terms of the integral basis $\{e_I\}_{I=1}^{22}$. Let us then define χ in terms of four integers n^a and m^a, $a = 1, 2$, as follows

$$\chi = (n^1 - \tau m^1) \alpha_1 + (n^2 - \tau m^2) \alpha_2. \tag{4.5.44}$$

Taking into account the self-duality of $\alpha_{1,2}$ and the fact that $\int_{K3} \alpha^1 \wedge \alpha^1 = \int_{K3} \alpha^2 \wedge \alpha^2 = 2$, the condition (4.5.40) reduces to

$$24 \,\mathrm{Im}\,\tau = L_{T^2}^2 \left(2|n^1 - \tau m^1|^2 + 2|n^2 - \tau m^2|^2 \right), \tag{4.5.45}$$

which, for fixed quantized numbers n^a and m^a, relates τ and $L_{T^2}^2$. Furthermore, in order to have both $\chi^{2,0}$ and $\chi^{0,2}$ non-zero, and hence $\mathcal{N} = 0$ supersymmetry, one needs to impose

$$\frac{n^2 - \tau m^2}{n^1 - \tau m^1} \neq \pm i. \tag{4.5.46}$$

Example 1

A particularly simple example is obtained by setting $n^1 = 1$ and $n^2 = m^1 = m^2 = 0$. In this case, we have a non-trivial fibration of the S^1 described by the coordinate x^1 introduced in (4.5.4) only, while the second S^1 described by x^2 is trivially fibered. The condition (4.5.45) then gives

$$L_{T^2}^2 = 12 \operatorname{Im} \tau, \tag{4.5.47}$$

while the H-field quantization conditions (4.5.37) reduce to

$$L_{T^2}^2 = k_1 \operatorname{Im} \tau, \qquad L_{T^2}^2 \operatorname{Re} \tau = k_2 \operatorname{Im} \tau, \tag{4.5.48}$$

with $k_{1,2} \in \mathbb{Z}$. From this one can infer a relation of k_1 and k_2

$$k_1 \operatorname{Re} \tau = k_2 . \tag{4.5.49}$$

If for example $\operatorname{Re} \tau = 0$, then $\tau = iR_2/R_1$ and $L_{T^2}^2 = R_1 R_2/\alpha'$. The condition (4.5.48) imposes that $k_1 = 12$ and $k_2 = 0$, while (4.5.47) provides the following constraint on the radius R_1

$$\frac{\alpha'}{R_1^2} \simeq \frac{1}{12} \simeq 0.083 . \tag{4.5.50}$$

Then $\frac{\alpha'}{R_1^2}$ is relatively small, moderately justifying the supergravity approximation. On the other hand, R_2 is obviously non-constrained. Notice that the profile of e^D is determined by (4.5.24). Since we are assuming that $N_{\text{NS5}} = 0$, for $L_{\text{K3}} \gg 1$ we can reasonably approximate $e^D \simeq 1$. Then, the formula (4.5.16) for the gravitino mass gives

$$m_{3/2} \simeq \frac{g_s M_P}{2 L_{\text{K3}}^4} \sqrt{\frac{R_1}{\pi R_2}} . \tag{4.5.51}$$

Example 2

To obtain a non-trivial fibration also of the second circle, let us take for example $n^1 = m^2 = 1$, $m^1 = n^2 = 0$ and $|\tau| = 1$ (with $|\operatorname{Re}\tau| \leq 1/2$), so that $R_1 = R_2$. Then (4.5.45) gives

$$L_{T^2}^2 = 6 \operatorname{Im} \tau, \tag{4.5.52}$$

while the flux quantization condition (4.5.37) still takes the form (4.5.48).

Clearly, the general solution of these equations is given by $k_1 = 6$ and $k_2 = 0, \pm 1, \pm 2, \pm 3$. By setting $k_2 = 0$, we get $R_1^2 = R_2^2 = \alpha'/6$. However, in this case $\operatorname{Re} \tau = 0$ and then $\chi^{0,2} = 0$, which implies that the solution is actually $\mathcal{N} = 1$. In the other cases $k_2 = \pm 1, \pm 2, \pm 3$, one gets $\mathcal{N} = 0$ solutions. For example, in the case of $k_2 = 1$ one gets

$$\operatorname{Re} \tau = \frac{1}{6}, \qquad \operatorname{Im} \tau = \frac{\sqrt{35}}{6}, \tag{4.5.53}$$

and
$$\frac{\alpha'}{R_1^2} = \frac{\alpha'}{R_2^2} = \frac{1}{\sqrt{35}} \simeq 0.17, \qquad (4.5.54)$$

which is even less relatively small than (4.5.50), and then even more moderately justifies the supergravity approximation. The gravitino mass (4.5.16) for these parameters is

$$m_{3/2} \simeq \frac{g_s M_P}{2\sqrt{\pi} L_{K3}^4} \times \frac{1}{6}, \qquad (4.5.55)$$

where the extra suppressing factor with respect to (4.5.51) with $R_1 = R_2$ comes from $\|\chi^{0,2}\| \simeq 1/6$.

These two simple examples should be sufficient to see how $\frac{1}{2}$DWSB vacua on an elliptically fibered K3 space can be constructed. It is however interesting too see already in this simple examples that the flux quantization condition (4.5.37) together with the integrated Bianchi identity (4.5.40) combine in order to restrict the size of the fiber. While on the one hand this leads to fiber sizes that are quite small and hence shed doubt on the justification of our supergravity approach, on the other hand this restriction in size guarantees that the base is always larger than the fiber, which is the essential condition for our whole approximation and for the mildness of supersymmetry breaking. Therefore, we believe that DWSB is a fruitful alternative to the type of SUSY-breaking normally discussed in the context of heterotic supergravity, which is gaugino condensation. In order to see the similarities and differences in the two approaches, we will in the next section analyze how an additional gaugino condensate affects our model.

4.6 Adding a gaugino condensate

Up to now we have focused on four-dimensional $\mathcal{N} = 0$ Minkowski vacua where the SUSY-breaking mechanism is due to the torsional geometry of the background. However, in the context of no-scale heterotic string compactifications, the source of supersymmetry breaking has traditionally been identified with the presence of a gaugino condensate generated by non-perturbative effects [173, 174]. It is therefore natural to incorporate a gaugino condensate to the above class of constructions, in order to see which new patterns of supersymmetry breaking it may lead to. In fact, since a gaugino condensate will modify the four-dimensional no-scale scalar potential, one may wonder whether its presence may restore supersymmetry and trigger the decay of the $\mathcal{N} = 0$ vacua discussed above to supersymmetric AdS_4 vacua.

4.6.1 Gaugino condensate and no-scale SUSY-breaking

A simple way to measure the effect of a gaugino condensate in a heterotic compactification is to incorporate the gaugino field up to quartic order into the supergravity action. In particular, one finds that the ten-dimensional string frame bosonic action is modified to

$$S = \frac{1}{2\kappa^2} \int d^{10}x \sqrt{-g}\, e^{-2\Phi} \left[\mathcal{R} + 4\,(d\Phi)^2 - \frac{1}{2}T^2 + \frac{\alpha'}{4}\,\mathrm{Tr}(R_+^2 - F^2 - 2\bar{\chi}\slashed{D}\chi)\right], \qquad (4.6.1)$$

where χ is the ten-dimensional gaugino field and we have defined the three-forms

$$T = H - \frac{1}{2}\Sigma, \qquad \Sigma_{MNP} = \frac{\alpha'}{4}\mathrm{Tr}\,\bar{\chi}\Gamma_{MNP}\chi\,. \qquad (4.6.2)$$

Let us now consider compactifications to four-dimensions, in which Σ (and therefore also T) has only internal legs. As already pointed out in [138, 173, 174] the presence of a non-trivial Σ modifies the scalar potential of the compactification. Indeed, following the computations of section 2.4.2, we see that the potential (4.1.6) is modified to

$$V' = V(H \to T) + \frac{\alpha'}{4\kappa_{10}^2} \int \mathrm{dvol}_M\, e^{4A-2\Phi}\, \bar{\chi}\slashed{D}_J\chi\,, \qquad (4.6.3)$$

where V is given by (4.1.6) with H substituted by T, and

$$\slashed{D}_J = \slashed{D} + \frac{1}{24}\,e^{-4A+2\Phi}[*\,d(e^{4A-2\Phi}J)]_{ijk}\Gamma^{ijk}\,. \qquad (4.6.4)$$

In order to get a four-dimensional Minkowski vacuum in this context, we again need to impose that V' is vanishing and extremized. By separately imposing that $V(H \to T)$ is extremized, one is naturally lead to consider configurations of the kind discussed in subsection (4.2.1), up to the replacement $H \to T$. Namely, one should impose

$$e^{2\Phi}d(e^{-2\Phi}J) = *T \qquad (4.6.5)$$

instead of (4.2.1b), and leave $dA = 0$, (4.2.1a), (4.2.8), and (4.2.28) unchanged. Furthermore, the gaugino term

$$\int \mathrm{dvol}_M\, e^{4A-2\Phi}\, \bar{\chi}\slashed{D}_J\chi \qquad (4.6.6)$$

must also be extremized. This leads to a set of conditions to be satisfied by χ and Σ.

Note that even in the case in which $W_1 = 0$ supersymmetry is still broken by the gaugino condensate, as one can check by looking directly at the supersymmetry transformations

$$\delta_\epsilon \Psi_I = \left(\nabla_I - \frac{1}{4}\mathcal{T}_I\right)\epsilon - \frac{1}{16}\Gamma_I \slashed{\Sigma}\epsilon, \tag{4.6.7a}$$

$$\delta_\epsilon \lambda = \left(\slashed{\partial}\Phi - \frac{1}{2}\mathcal{T}\right)\epsilon - \frac{3}{8}\slashed{\Sigma}\epsilon, \tag{4.6.7b}$$

$$\delta_\epsilon \chi = \frac{1}{2}\slashed{F}\epsilon. \tag{4.6.7c}$$

In particular, for compactifications to flat space and non-vanishing gaugino condensate, the external gravitino variation is always non-vanishing, as $\delta\Psi_\mu = -\frac{1}{16}\Gamma_\mu\slashed{\Sigma}\epsilon \neq 0$. One may then restore supersymmetry by considering compactifications to AdS$_4$, as analyzed in [198]. Such kind of compactifications will be considered in the next subsection.

As discussed in section 4.3 in the absence of a gaugino condensate the background condition (4.1.9c) can be interpreted in terms of calibrations for gauge bundles and space-time filling NS5-branes. Remarkably, the modification of (4.1.9c) into (4.6.5) is exactly the necessary one in order to preserve this interpretation. As in section 4.3 this can be seen by going to the dual formulation, briefly reviewed in appendix A.2, where one uses the seven-form $\hat{H}$ instead of H as fundamental field. Recall that $\hat{H}$ is the flux which couples electrically to NS5-branes, and hence the one to appear in the generalized calibration. As discussed in appendix A.2 in the presence of a gaugino condensate $\hat{H}$ and H are related by $\hat{H} = e^{-2\Phi} *_{10} T = *_{10}(H - \frac{1}{2}\Sigma)$. We can then split $\hat{H}$ as

$$\hat{H} = \text{dvol}_{X_4} \wedge \tilde{H} = \text{dvol}_{X_4} \wedge (e^{4A-2\Phi} * T). \tag{4.6.8}$$

Note that (4.6.5) ensures that $\tilde{H}$ is closed and even exact, and so we can write $\tilde{H} = d\tilde{B}$, where $\tilde{B}$ is an internal potential two-form.

It is in fact illustrating to express the full potential (4.6.3) in the dual formulation. Starting from the dual action given in (A.2.6), one arrives at the potential

$$\tilde{V}' = V(H \to -e^{-4A+2\Phi} * \tilde{H}) + \frac{\alpha'}{4\kappa_{10}^2} \int \text{dvol}_M\, e^{4A-2\Phi} \overline{\chi}\slashed{D}_J \chi$$
$$- \frac{1}{2\kappa_{10}^2}\int_M (e^{4A-2\Phi}J - \tilde{B}) \wedge [dH - \frac{\alpha'}{4}(\text{Tr}\, R_+ \wedge R_+ + \text{Tr}\, F \wedge F)], \tag{4.6.9}$$

where the potential V has again the form (4.1.6). We see that the DWSB ansatz, supplemented by the Bianchi identity (2.2.5) and the extremization of (4.6.6) is sufficient to get a vacuum, since the last term in (4.6.9) can be seen as being quadratic in vanishing terms because of (2.2.5) and (4.6.5). Note that in this formulation $\tilde{H}$ and χ are regarded as independent fields and that this gives a simple interpretation of the no-scale structure observed in [174]. Indeed, by starting from a Calabi-Yau compactification, one can allow a non-trivial gaugino condensate $\Sigma \neq 0$ by taking

χ such that $\slashed{D}_{CY}\chi = 0$ and still imposing $\tilde{H} = 0$. Of course, by going back to the ordinary formulation, the latter translates into $H = \frac{1}{2}\Sigma \neq 0$, as originally found in [174].

Let us stress that so far χ has not been restricted at all. Of course, χ should allow for a 4d + 6d splitting $\chi = \chi_{4D} \otimes \chi_{6D} + $ c.c., with χ_{4D} playing the role of the condensing gaugino in four-dimensions. Also, χ_{6D} (and thus Σ) cannot be completely arbitrary, but should obey certain consistency conditions, like for instance those derived from the potential (4.6.6), and the other set of equations that must be imposed on the background. In particular note that by imposing (4.6.5), we have

$$\slashed{D}_J|_{(4.6.5)} = \slashed{D}_T = \slashed{D} - \frac{1}{4}\slashed{T}, \qquad (4.6.10)$$

where $\slashed{D}_T$ is the Dirac operator for the gaugino, cf. (A.2.6)

$$S_{\text{gaugino}} = -\frac{\alpha'}{4\kappa_{10}^2}\int d^{10}x\sqrt{-g}\, e^{-2\Phi}\,\bar{\chi}\slashed{D}_T\chi\,. \qquad (4.6.11)$$

This means that if we impose the DWSB conditions of section 4.2 together with the gaugino equations of motion on our background, then the full potential (4.6.3) vanishes, consistently with the requirement of having a four-dimensional Minkowski vacuum.

As a subset of the above class of vacua one may consider the case in which we have a torsional but complex manifold M. This implies that $W_1 = 0$, and so supersymmetry is not broken at the classical level as in section 4.2, but just by the presence of the gaugino condensate. This generalization to torsional backgrounds of the Calabi-Yau models considered in [174] has been proposed in [138] as a way to achieve a richer pattern of moduli stabilization and supersymmetry. As argued there, the fact that M is complex together with the Bianchi identity implies the choice $\chi_{6D} = \eta_+$, where η_+ comes from the 6d component of the Killing spinor ϵ of the compactification, split as in (4.1.1). Then $\Sigma = \Sigma^{3,0} + \Sigma^{0,3}$ and (4.2.1b) can be written as

$$e^{2\Phi}\partial(e^{-2\Phi}J) = i\,H^{2,1}, \qquad H^{3,0} = \frac{1}{2}\Sigma^{3,0}, \qquad (4.6.12)$$

so the $(2,1)$ component of H is naturally associated with the compactification scale, while the $(3,0)$ component is associated to the, presumably lower, gaugino condensate scale. Moreover, we have that

$$\bar{\chi}\slashed{D}_J\chi \sim \Omega \cdot dJ|_{W_1=0} + \text{c.c.} = 0\,. \qquad (4.6.13)$$

Hence, the second piece in (4.6.3) also vanishes.

The two different scales associated to the components $H^{2,1}$ and $H^{3,0}$ of the H-flux suggest that, in principle, below the scale of $H^{2,1}$ one could truncate the potential (4.6.3) by imposing the first equation in (4.6.12). In general this would imply freezing the vevs of several compactification moduli in such a way that the first equation in (4.6.12) is satisfied. We would then be left with a truncated potential of the form

$$V_{\text{no-scale}} = \frac{1}{4\kappa^2} \int_M \text{dvol}_M e^{4A-2\Phi} (H^{3,0} + H^{0,3} - \frac{1}{2}\Sigma)^2, \quad (4.6.14)$$

which has exactly the form of the no-scale potential considered in [174]. It is not clear, however, whether this no-scale structure will survive at the scale set by $H^{2,1}$, since at this scale we may change the vevs of the complex structure moduli, which in turn change the definition of $H^{3,0}$.

4.6.2 Supersymmetric AdS_4 vacua and calibrations

As recalled above, Σ enters the supersymmetry transformations (4.6.7) in such a way that it always breaks supersymmetry in compactifications to Minkowski space. Indeed, by taking a metric ansatz of the form (2.4.15) and following the computations in [198] one finds that supersymmetry requires a non-vanishing cosmological constant and allows for a possible non-trivial warping, in sharp contrast with the perturbative results of section 2.4.2. More precisely, one defines the AdS_4 Killing spinor ζ as $\nabla_\mu \zeta = \frac{1}{2}\overline{w}_0 \hat{\gamma}_\mu \zeta^*$, where w_0 is a constant related to the AdS_4 radius by

$$|w_0|^2 = \frac{1}{R_{\text{AdS}}^2}. \quad (4.6.15)$$

Then, the external gravitino supersymmetry requires that

$$\Omega \lrcorner \Sigma = 8 e^{-A} w_0, \qquad dA = -\frac{1}{8} * (J \wedge \Sigma), \quad (4.6.16)$$

which indeed reduce to the results of section 2.4.2 for $\Sigma = 0$.

Moreover, a non-vanishing Σ will modify the supersymmetry conditions (4.1.9). Following again the computations in [198] it is easy to see that the remaining supersymmetry conditions can be rewritten as

$$e^{-4A+2\Phi} d\left(e^{4A-2\Phi} J\right) = *T + 3 e^{-A} \text{Im}\left(\overline{w}_0 \Omega\right), \quad (4.6.17a)$$
$$e^{-3A+2\Phi} d\left(e^{3A-2\Phi} \Omega\right) = -w_0 e^{-A} J \wedge J, \quad (4.6.17b)$$

which again reduces to (4.1.9) for $w_0 = dA = 0$. In fact, (4.6.17b) implies that

$$d\left(e^{2A-2\Phi} J \wedge J\right) = 0, \quad (4.6.18)$$

generalizing equation (4.1.9b) for non-constant warping. Finally, it is easy to see that the equations

(4.6.17) imply that

$$*T = -\frac{3}{2}e^{-A}\text{Im}\,(\overline{w}_0\Omega) + \text{d}(3A - \Phi) \wedge J + W_3\,. \quad (4.6.19)$$

What happens then to the potential (4.6.3) when the above set of supersymmetry conditions are imposed? By plugging (4.6.17) into the first term on the right hand side of (4.6.3) one gets[28]

$$\begin{aligned}V(H \to T)|_{\text{SUSY}} &= \frac{3}{\kappa_{10}^2}\int \text{dvol}_M\, e^{2A-2\Phi}\left\{9|\text{Im}(\overline{w}_0\Omega)|^2 + |w_0|^2\big(|J \wedge J|^2 - |J \wedge J \wedge J|^2\big)\right\}\\ &= \frac{3|w_0|^2}{\kappa^2}\int \text{dvol}_M\, e^{2A-2\Phi}\,. \end{aligned} \quad (4.6.20)$$

On the other hand, by also imposing the gaugino equations of motion derived from (4.6.11), one obtains that the second term in (4.6.3) gives

$$\begin{aligned}\frac{\alpha'}{4\kappa_{10}^2}\int \text{dvol}_M\, e^{4A-2\Phi}\,\overline{\chi}\slashed{D}_J\chi|_{\text{SUSY}} &= -\frac{3\alpha'}{16\kappa_{10}^2}\int \text{dvol}_M\, e^{3A-2\Phi}\overline{\chi}\text{Re}(\overline{w}_0\slashed{\Omega})\chi \\ &= -\frac{3}{4\kappa_{10}^2}\int \text{dvol}_M\, e^{3A-2\Phi}\text{Re}(\overline{w}_0\Omega)\lrcorner\Sigma \\ &= -\frac{6|w_0|^2}{\kappa_{10}^2}\int \text{dvol}_M\, e^{2A-2\Phi}\,,\end{aligned} \quad (4.6.21)$$

where in the last step we have used the first of (4.6.16). Note that the gaugino equations of motion are automatically satisfied if we decompose the ten-dimensional gaugino as $\chi = \chi_{4D} \otimes \eta_+ + \text{c.c.}$. Combining (4.6.20) and (4.6.21) and using (2.4.32), we get

$$V'|_{\text{SUSY}} = -\frac{3|w_0|^2}{\kappa_{10}^2}\int e^{2A-2\Phi}\text{dvol}_M = -\frac{3M_P^2}{R_{\text{AdS}}^2}\,. \quad (4.6.22)$$

Hence, we reproduce the expected value of the potential energy of an AdS$_4$ compactification with cosmological constant $\Lambda_{\text{AdS}} = -3/R_{\text{AdS}}^2$. Note that the contribution from the gaugino term (4.6.21) is crucial to get the correct result. On the other hand, note that in order to evaluate this contribution we have imposed the equations of motion of the gaugino. It seems technically difficult to do it otherwise and so, unlike for purely bosonic backgrounds, we do not have a direct off-shell expression for the scalar potential.

Despite this, one can still analyze the supersymmetry conditions of backgrounds with fermion condensates and interpret them in terms of calibrations. In particular, by direct comparison with equations (4.1.9) and the discussion in sections 4.3, one would still expect the following dictionary between calibrations and BPS-objects of the compactification

---

[28]Here, we assume that the deviation from the flat supersymmetric case is at least of order α'. Hence, we neglect contributions to the potential energy coming from the curvature and $\mathcal{O}[(\nabla A)^2]$ terms in (4.1.6b).

Calibration	10d BPS object	4d BPS object
$e^{4A-2\Phi}J$	NS5 on $X_4 \times \Pi_2$	gauge theory
$e^{3A-2\Phi}\Omega$	NS5 on $X_3 \times \Pi_3$	domain wall
$e^{2A-2\Phi}J \wedge J$	NS5 on $X_2 \times \Pi_4$	string

with again Π_p a p-dimensional submanifold of M and X_d a d-dimensional slice of X_4. By extending the results of [223] to NS5-branes, one can check that (4.6.17) and (4.6.18) indeed correspond to the existence of generalized calibrations for NS5-branes in an AdS$_4$ background.

4.6.3 $\frac{1}{2}$DWSB AdS$_4$ vacua with gaugino condensate

Having understood in terms of calibrations the conditions for heterotic four-dimensional $\mathcal{N}=1$ AdS$_4$ vacua with a gaugino condensate, it is now clear how to implement our previous strategy to construct $\mathcal{N}=0$ AdS$_4$ backgrounds of the same sort. Recall from section 4.2 that for the $\mathcal{N}=0$ vacua considered there the supersymmetry conditions (4.2.1) were still satisfied, allowing to define a stable gauge bundle as in section 4.3. On the other hand, the domain-wall BPSness condition was relaxed to (4.2.2), and half-imposed for $\frac{1}{2}$DWSB backgrounds via (4.4.1).

In the case of AdS$_4$ compactifications, the surviving 1/2 domain-wall BPSness is determined by (4.6.17a) itself, since the equation of motion $d(e^{4A-2\Phi} * T) = 0$ for T implies that

$$d[e^{3A-2\Phi}\text{Im}(\overline{w}_0 \Omega)] = 0 \ . \quad (4.6.23)$$

We are thus naturally led to consider $\mathcal{N}=0$ AdS$_4$ backgrounds, where supersymmetry breaking originates from a background condition of the form[29]

$$e^{-3A+2\Phi}d\big[e^{3A-2\Phi}\text{Re}(\overline{w}_0 \Omega)\big] = \text{Re}(\overline{w}_0 W_1) J \wedge J + \text{Re}(\overline{w}_0 W_2) \wedge J \ , \quad (4.6.24)$$

so that supersymmetry is broken when $\text{Re}(\overline{w}_0 W_1) \neq -e^{-A}|w_0|^2$ or $W_2 \neq 0$. This implies that these backgrounds are characterized by the torsion classes of the internal manifold M via

$$W_4 = d(\Phi - A) \ , \qquad \text{Re}\, W_5 = d\Phi - \frac{3}{2}dA \ , \qquad \text{Im}(\overline{w}_0 W_1) = \text{Im}(\overline{w}_0 W_2) = 0 \ , \quad (4.6.25)$$

while W_3 is specified by the relation

$$*T = -\frac{3}{2}\text{Im}\big((\overline{W}_1 + 2e^{-A}\overline{w}_0)\Omega\big) + d(3A - \Phi) \wedge J + W_3 \ . \quad (4.6.26)$$

---

[29]That this form is consistent with our ansatz can be shown as in section 4.2.1. The actual computation is given in appendix A.3.

A further source of supersymmetry breaking comes from the external gravitino supersymmetry conditions. More precisely, setting

$$\Omega \lrcorner \Sigma \;=\; 8\, e^{-A} \sigma_0 \qquad (4.6.27)$$

one finds that the external gravitino supersymmetry is broken if $\sigma_0 \neq w_0$. Hence, we have two natural SUSY-breaking scalar parameters for this kind of compactifications

$$\mathcal{I}_1 \;=\; W_1 + e^{-A} w_0\,, \qquad \mathcal{I}_2 \;=\; \frac{1}{2} e^{-A} (\sigma_0 - w_0)\,. \qquad (4.6.28)$$

Note that this family of backgrounds contains all of the supergravity compactifications with non-vanishing Σ considered up to date in the literature. In particular, it generalizes the compactifications analyzed in [138, 174, 244], where vacua with $W_1 = w_0 = 0$ have been discussed, and where thus only the SUSY-breaking parameter $\mathcal{I}_2$ was turned on. Such $\mathcal{N} = 0$ constructions are in some sense orthogonal to the ones considered in section 4.2, since there we had $\sigma_0 = w_0 = 0$ and so only $\mathcal{I}_1 \neq 0$. It is therefore natural to wonder to what extent four-dimensional vacua for arbitrary values of both SUSY-breaking parameters turned on can be constructed.

As before, some amount of information can be obtained by analyzing the scalar potential (4.6.3). One can easily see that the first term on the right hand side of (4.6.3) reads

$$V(H \to T)|_{\frac{1}{2}\text{DWSB}} \;=\; \frac{1}{4\kappa_{10}^2} \int e^{2A - 2\Phi} \text{dvol}_M \left(36 |w_0|^2 + |W_2|^2 - 24 |W_1|^2 \right), \qquad (4.6.29)$$

while, by imposing the gaugino equations of motion derived from (4.6.11) and following the same steps as in (4.6.21), the second term on the right hand side of (4.6.3) gives

$$\frac{\alpha'}{4\kappa_{10}^2} \int \text{dvol}_M\, e^{4A - 2\Phi}\, \overline{\chi} \slashed{D}_J \chi |_{\frac{1}{2}\text{DWSB}} \;=\; -\frac{6}{\kappa_{10}^2} \int \text{dvol}_M\, e^{2A - 2\Phi}\, \text{Re}(\overline{w}_0 \sigma_0)\,. \qquad (4.6.30)$$

Summing up these two terms one gets

$$V'|_{\frac{1}{2}\text{DWSB}} \;=\; \frac{1}{4\kappa_{10}^2} \int e^{2A - 2\Phi} \text{dvol}_M \left[36 |w_0|^2 - 24 |W_1|^2 + |W_2|^2 - 24 \text{Re}(\overline{w}_0 \sigma_0) \right], \qquad (4.6.31)$$

which on-shell should equal $-3|w_0|^2 M_P^2$. This is indeed the case for unbroken SUSY, as already considered in section 4.6.2, since there we have that $W_2 = \mathcal{I}_1 = \mathcal{I}_2 = 0$, which in turn implies that $w_0 = \sigma_0 = -e^A W_1$.

If on the other hand we consider the torsional geometries considered in section 4.2, which are associated to Minkowski DWSB vacua, we need to impose the constraint $|W_2|^2 = 24|W_1|^2$ on the above vacuum energy. One then concludes that, by consistency, an AdS$_4$ vacuum of this kind needs to satisfy the relation $\sigma_0 = 2w_0$,[30] and so supersymmetry is necessarily broken because the first equation in (4.6.16) is not satisfied. This fact shows that, naively, adding a gaugino condensate on top of the $\mathcal{N} = 0$ torsional geometries of section 4.2 is not enough to restore the supersymmetry of the compactification. Indeed, from section 4.6.2 we see that in order to construct supersymmetric AdS$_4$ vacua the torsion class W_2 must vanish. By equation (4.2.8), this is not possible for the Minkowski DWSB vacua of section 4.2, since there by assumption $W_1 \neq 0$. In particular, if we consider the fibered manifolds of section 4.4 we see that M should undergo some kind of topology change in order to flow to a manifold M' with $W_2 = 0$. It is not clear how the presence of Σ could trigger such topology change, so one would expect that adding a gaugino condensate on top of the no-scale vacua of section 4.4 would most likely take them to a $\frac{1}{2}$DWSB AdS$_4$ background of the kind considered here, not being clear if this would be a vacuum of the theory. Of course, adding further non-perturbative effects produced by, e.g. worldsheet instantons may provide the necessary ingredients to promote our heterotic no-scale vacua to an $\mathcal{N} = 1$ AdS$_4$ vacuum, along the lines of [245].

This ends our discussion of DWSB in the context of heterotic string theory. We want to point out again the crucial role that the use of SU(3) structures had in our construction. Without them even to reach at our starting point, namely the scalar potential (4.1.6) of BPS-like form, would have been impossible. One can trace this back to the fact that the underlying SU(3) structure of the manifold makes it possible to rewrite the curvature scalar of the internal manifold and the SUSY variation of the gravitino, which are two highly non-trivial expressions, in a very simple and suggestive form. With this firm starting point it was then comparably easy to find conditions for a mild, controllable SUSY-breaking.

Again, the notion of SU(3) structure was very useful in our discussion of calibrations which provided a physical explanation to our, at first sight, ad hoc ansatz. However, as we tried to explain, the necessity to be still able to define a gauge bundle in a well defined way makes our choice of SUSY-breaking very natural. With the tools given at hand by SU(3) structures we were finally able to find a subclass of our ansatz for which it was possible to construct consistent non-supersymmetric vacua of heterotic supergravity.

The inclusion of a gaugino condensate gave a nice completion of the topic, as it brings together our new ansatz and the well known case of SUSY-breaking via a fermionic condensate. Besides documenting the changes that our formalism undergoes when such a condensate is added, we were also able to show that it is not possible to reach a supersymmetric configuration by including a

---

[30]In fact, one could also consider the case where $\sigma_0 = 2w_0 + biw_0$, $b \in \mathbb{R}$. However, σ_0 should enter the holomorphic gravitino mass that can be calculated following section 4.2.2. Since the gravitino mass is proportional to the superpotential and the latter is expected to be aligned with the phase of w_0 (as the supersymmetric case shows), we are naturally led to expect that σ_0 has the same phase as w_0.

gaugino condensate within our ansatz.

However, one should note at the end of this section that our whole discussion has taken place in the regime of small string coupling g_s. As was shown by Witten in [91] this regime is not very suitable to reach realistic values for the various coupling constants of nature in a consistent way. It is therefore an interesting question whether our ansatz can be lifted to the regime of strong coupling, i.e. whether it is possible to implement DWSB in heterotic M-theory. We tried to provide the first steps in answering this question in [4] and will present our findings in the next chapter.

Chapter 5

BPS-Potentials in M-Theory

We saw in the last chapter how non-supersymmetric vacua can be constructed for weakly coupled heterotic string theory in the low energy regime, i.e. for heterotic supergravity. The strong coupling behavior of these non-superymmetric theories should most likely be describable in terms of Horava-Witten theory if the DWSB ansatz that we employed in the last chapter could be lifted to the M-theory setting.

In order to make this possible, one has to satisfy the following main conditions. Firstly, it should be possible to reach the weak coupling limit by dimensionally reducing the eleven-dimensional theory. Since in this limit we deal with a six-dimensional internal manifold with SU(3) structure, one should start in eleven dimensions by compactifying on a seven-dimensional manifold that also has SU(3) structure. As we discussed in chapter 3 these manifolds can be viewed locally as the direct product of a six-dimensional SU(3) structure submanifold and an additional seventh dimension, which serves naturally as the direction of dimensional reduction. We will therefore concern ourselves with compactifications on seven-dimensional SU(3) structure manifolds.

Secondly, it should be possible to define a BPS-like potential also in the case of heterotic M-theory. To understand this we should once more repeat the logic of the last chapter. There, we started with a BPS-like action in order to be sure that any supersymmetric vacuum satisfies the equations of motion. Only being sure of this it was possible to deform the SUSY conditions such that the DWSB ansatz emerges. Since then the potential was not any longer automatically extremized, we also had to deal with an additional equation of motion, which restricted the geometry so severe (at least in the case of $\frac{1}{2}$DWSB) that, in the end, it was possible to find explicit solutions. But without the BPS-like form of the potential, this approach would have been void, since there would have been no guarantee that the EoM's are satisfied from the beginning. One thus sees that also in the M-theory case one should first rewrite the potential[1]

$$V_0 = \frac{1}{2\kappa^2}\int_M \text{dvol}_7 \left[-e^{2A}\hat{R}^{(4)} - e^{4A}(R - 8\nabla^2 A - 20\,\text{d}A^2 - \frac{1}{2}|G|^2 - \frac{1}{2}\mu^2 - \mu\, C \lrcorner * G)\right],$$

$$V_b = \frac{1}{8\pi\kappa^2}\left(\frac{\kappa}{4\pi}\right)^{2/3}\sum_{p=1,2}\int_{B_{6,p}} \text{dvol}'_{6,p}\, e^{4A'-3\sigma}\left\{\left(\text{Tr}(\mathcal{F}^{(6)}_p)^2 - \frac{1}{2}\text{Tr}(R^{(6)'}_{p+})^2\right)\right.$$
$$\left. - \frac{1}{24}\left|e^{-2A'}\hat{R}^{(4)} - 12|\text{d}A'|^2\right|^2 - 4\,e^{-2A'}(\nabla_i\nabla_j e^{A'})(\nabla^i\nabla^j e^{A'}) - 2\left|\text{d}A'\lrcorner H\right|^2\right\}$$

in terms of the supersymmetry conditions, and check whether or not it has a BPS-like form.

This is essentially what we have done in [4] and what we will present in this chapter. The main obstacle in this work was the rewriting of the curvature scalar of the internal manifold in terms of the SU(3) structure forms of our seven-dimensional compactification manifold. Since every seven-dimensional manifold, which has a globally well defined spinor, has also a G_2 structure, and since an expression for the curvature scalar is known for this case, we considered the problem of a G_2 manifold first and then specialized to the SU(3) structure case. As it turned out the expression is quite involved and it is difficult to implement the SUSY-conditions into it. We therefore give a detailed description of the supersymmetry conditions and of our method of bringing them together with the expression for the curvature scalar.

Unfortunately, our findings are that for a general seven-dimensional SU(3) structure manifold it is not possible to bring the potential into a BPS-like form. Since this result is quite unexpected, we provide several simple limits with known outcomes to confirm its consistency. We then also comment whether or not a lift of the models discussed in chapter 4 is possible.

5.1 The Ricci scalar of G_2 manifolds

As we pointed out in the introduction to this chapter a main obstacle in the analysis of (2.4.8) is the seven-dimensional Ricci scalar R, as it is not possible to see how this quantity behaves for broken or unbroken SUSY. Thus, its standard form is not suitable for the discussion of the BPSness of the potential. As in chapter 4 we will therefore use that all information encoded in the metric g of a G-structure manifold is also contained in the forms invariant under G, and express the Ricci scalar in terms of these forms. We start with the discussion of the curvature scalar of manifolds with G_2 structure and derive from this an expression in terms of SU(3) structure.

---

[1]Of course, if not stated otherwise all quantities appearing in this chapter belong to the eleven-dimensional theory and its compactification to four dimensions described in section 2.1 and section 2.4.1.

5.1.1 R in terms of G_2 structure

The Ricci scalar for G_2 structure manifolds was worked out in [210] in terms of the G_2 torsion classes defined by (3.3.3) and reads

$$R = -12 * d * \tau_1 + \frac{21}{8} \tau_0^2 + 30 |\tau_1|^2 - \frac{1}{2} |\tau_2|^2 - \frac{1}{2} |\tau_3|^2 . \tag{5.1.1}$$

Due to the complicated dependence of τ_2 and τ_3 on ϕ and ψ given in (3.3.5) this seems not to be a very pleasing formula. But as we show in appendix A.4, it is possible to use the results of [210] in order to connect the absolute values of the torsion classes with each other

$$\begin{aligned} |\tau_2|^2 &= |d\psi|^2 - 48 |\tau_1|^2 , \\ |\tau_3|^2 &= |d\phi|^2 - 36 |\tau_1|^2 - 7 |\tau_0|^2 . \end{aligned} \tag{5.1.2}$$

Using these equations, the scalar curvature R can be rewritten in terms of ϕ and ψ in a quite suggestive way

$$\begin{aligned} R &= -12 * d * \tau_1 + \frac{49}{8} \tau_0^2 + 72 |\tau_1|^2 - \frac{1}{2} |d\phi|^2 - \frac{1}{2} |d\psi|^2 \\ &= -\nabla^m (d\psi \lrcorner \psi)_m + \frac{1}{2} |d\psi \lrcorner \psi|^2 + \frac{1}{8} |d\phi \lrcorner \psi|^2 - \frac{1}{2} |d\phi|^2 - \frac{1}{2} |d\psi|^2 , \end{aligned} \tag{5.1.3}$$

as it depends only on ϕ, ψ, and their exterior derivatives.

5.1.2 R in terms of SU(3) structure

Although equation (5.1.3) provides a good description of R as a function of ϕ and ψ, this form is not convenient for our purposes, as we are interested in manifolds with SU(3) structure. The next task is thus to decompose ϕ and ψ according to (3.3.16) and find the expression for R in terms of v, J, and Ω. A lengthy calculation provides the building blocks of R

$$\begin{aligned} |d\phi|^2 &= |v \wedge dJ - \text{Re} \, d\Omega|^2 - |dv \wedge J - \text{Re} \, d\Omega|^2 + |\text{Re} \, d\Omega|^2 + 2 |dv \wedge J|^2 \\ &+ |dv \lrcorner v - dJ \lrcorner J|^2 - |dv \lrcorner v|^2 - |dJ \lrcorner J|^2 , \end{aligned} \tag{5.1.4}$$

$$\begin{aligned} |d\psi|^2 &= \left|\frac{1}{2} dJ^2 + dv \wedge \text{Im} \, \Omega\right|^2 - |\text{Im} \, d\Omega \lrcorner v|^2 - |dv \lrcorner v|^2 - \left|\frac{1}{2} dJ^2 \lrcorner v\right|^2 \\ &+ |dv \lrcorner v + \text{Im} \, d\Omega \lrcorner \text{Im} \, \Omega|^2 - |\text{Im} \, d\Omega \lrcorner \text{Im} \, \Omega|^2 + \left|\frac{1}{2} dJ^2 \lrcorner v - \text{Im} \, d\Omega\right|^2 , \end{aligned} \tag{5.1.5}$$

$$\left|\mathrm{d}\psi\lrcorner\psi\right|^2 = \left|\frac{1}{4}\mathrm{d}J^2\lrcorner J^2 - \operatorname{Re}\mathrm{d}\Omega\lrcorner(v\wedge J) + \mathrm{d}v\lrcorner\operatorname{Im}\Omega - 2\,\mathrm{d}v\lrcorner v - \operatorname{Im}\mathrm{d}\Omega\lrcorner\operatorname{Im}\Omega\right|^2 \quad (5.1.6)$$
$$+ \frac{1}{4}\left|\mathrm{d}J^2\lrcorner(v\wedge J^2)\right|^2 - \left|\operatorname{Im}\mathrm{d}\Omega\lrcorner(v\wedge\operatorname{Im}\Omega)\right|^2 + \frac{1}{16}\left|\operatorname{Im}\mathrm{d}\Omega\lrcorner(J^2 + 2v\wedge\operatorname{Im}\Omega)\right|^2$$
$$+ \frac{1}{4}\left|\operatorname{Im}\mathrm{d}\Omega\lrcorner J^2\right|^2 - \frac{1}{16}\left|2\,\mathrm{d}J^2\lrcorner(v\wedge J^2) + \operatorname{Im}\mathrm{d}\Omega\lrcorner(J^2 + 2v\wedge\operatorname{Im}\Omega)\right|^2,$$

$$\left|\mathrm{d}\phi\lrcorner\psi\right|^2 = \left|2\mathrm{d}v\lrcorner J + \frac{1}{2}\operatorname{Re}\mathrm{d}\Omega\lrcorner J^2 - \mathrm{d}J\lrcorner\operatorname{Im}\Omega + \operatorname{Re}\mathrm{d}\Omega\lrcorner(v\wedge\operatorname{Im}\Omega)\right|^2 \quad (5.1.7)$$
$$= \left|2\mathrm{d}v\lrcorner J + \operatorname{Re}\mathrm{d}\Omega\lrcorner J^2 - \frac{1}{2}\operatorname{Im}[\mathrm{d}\Omega\lrcorner(v\wedge\bar{\Omega})]\right|^2,$$

$$\mathrm{d}\psi\lrcorner\psi = \frac{1}{4}\mathrm{d}J^2\lrcorner J^2 + \mathrm{d}v\lrcorner\operatorname{Im}\Omega - \frac{1}{2}v(\operatorname{Im}\mathrm{d}\Omega\lrcorner J^2) - \operatorname{Re}\mathrm{d}\Omega\lrcorner(v\wedge J) \quad (5.1.8)$$
$$- 2\mathrm{d}v\lrcorner v - \operatorname{Im}\mathrm{d}\Omega\lrcorner\operatorname{Im}\Omega - v\left(\operatorname{Im}\mathrm{d}\Omega\lrcorner(v\wedge\operatorname{Im}\Omega)\right),$$

which are at this stage not very illuminating.[2] However, we have reached our first goal, namely we have rewritten the Ricci scalar R purely in terms of the SU(3) structure forms v, J, and Ω. Note that most of the appearing parts are square terms, but there are also a lot of linear contributions coming from $\mathrm{d}\psi\lrcorner\psi$. In order to check for a BPS-like potential we have to know whether the square terms vanish when we impose supersymmetry. To this end we turn to the investigation of the SUSY conditions in the next section.

5.2 Supersymmetry conditions

The only fermionic fields appearing in Horava-Witten theory are the gravitino and the two gauginos living on the two boundaries. The supersymmetry variations of these fields are given in (2.1.4) and in (2.1.7), respectively, and have to vanish for a supersymmetric setting. As in chapter 4 one should decompose the Majorana spinor ϵ appearing in the SUSY variation in a four-dimensional and a seven-dimensional spinor, suitable for seven-dimensional SU(3) structure manifolds.

However, in [208] it was shown that there are three ways to decompose the spinor ϵ such that one obtains an SU(3) invariant SO(7) spinor, which can be identified with the spinor η_+ of section 3.3. The possibilities are further restricted to two if one wants to consider $\mathcal{N} = 1$ SUSY in four dimensions. In order to decide, which one of these two should be used in our case, one has to keep in mind that the spinor ϵ should reduce to its counterpart in ten dimensions after dimensional reduction. For $\mathcal{N} = 1$ SUSY this means that in type IIA there should be two internal spinors

---
[2] Note that we used $\mathrm{d}J^2\lrcorner(v\wedge\operatorname{Im}\Omega) = -2\operatorname{Re}\mathrm{d}\Omega\lrcorner(v\wedge J)$ during the calculation.

of opposite chirality, while in the heterotic case there can only be one chiral spinor. As it turns out only one of the possibilities of [208] can give the heterotic limit. The other $\mathcal{N} = 1$ case must therefore be associated to M-theory without boundaries. We will comment on this case briefly in appendix A.6.

Then, to describe heterotic M-theory one is restricted to use the following decomposition into a chiral 4d spinor χ_+ and the SU(3) structure spinor η_+

$$\epsilon = \chi_+ \otimes \eta_+ + \chi_- \otimes \eta_- = \chi_+ \otimes \eta_+ + c.c. \, . \tag{5.2.1}$$

Using this split the gaugino variations can be rewritten with the help of (3.3.11) and yield the conditions that $\mathcal{F}_p$ is $(1,1)$ and primitive with respect to the almost complex structure[3] defined by Ω and J

$$\mathcal{F}_p \lrcorner J = 0, \qquad \mathcal{F}_p^{2,0} = \mathcal{F}_p^{0,2} = 0 \, . \tag{5.2.2}$$

The eleven-dimensional gravitino variation gives rise to two sets of conditions[4]

$$\delta\Psi_\mu = 0 \;\Rightarrow\; e^{-A} w_0 \eta_+^* + \left(\slashed{\partial} A + \frac{1}{6}\slashed{G} + \frac{i\mu}{3}\right)\eta_+ = 0, \tag{5.2.3a}$$

$$\delta\Psi_m = 0 \;\Rightarrow\; \nabla_m \eta_+ = \frac{1}{288}\left(i\mu\gamma_m + 8\,G_{mpqr}\gamma^{pqr} - G_{npqr}\gamma_m{}^{npqr}\right)\eta_+ \, . \tag{5.2.3b}$$

The first of these equations will give algebraic constraints on the flux G_{11}. As is easy to see from (3.3.11) the contraction of (5.2.3a) with $\eta_+^\dagger$ leads to $\mu = 0$. Therefore, we will consider only internal four-flux and set $\mu = 0$ in what follows. The second equation can be translated into differential conditions on v, J, and Ω. Similar analyses have also been performed by [207, 208, 211–214], whose results are equivalent to ours.

5.2.1 Differential conditions

Contracting (5.2.3b) with $\eta_+^\dagger \gamma_{n_1...n_{p-1}}$ and anti-symmetrizing over all indices gives the exterior derivatives of Σ_p. Exchanging $\eta_+^\dagger$ with η_+^T yields the derivatives of $\tilde{\Sigma}_p$, which can be converted with (3.3.11) into the derivatives of v, J, Ω, and their wedge products. Furthermore, $dZ = dA$ and $d(v \wedge J) = dv \wedge J - v \wedge dJ$ demand that $w_0 = 0$. For supersymmetric vacua we are thus dealing with compactifications to warped Minkowski space, that obey only internal flux, and whose

---
[3]In this chapter we will often leave out the explicit reference to Ω and J, when we speak of (p,q)-forms. We remind the reader that in any case such statements are only valid locally on the six-dimensional subspace perpendicular to v.

[4]Here, we used the AdS killing spinor equation $\hat{\nabla}_\mu \chi_+ = 1/2\, w_0^* \hat{\gamma}_\mu \chi_-$.

internal manifold has to satisfy the conditions[5]

$$
\begin{align}
e^{-2A}\mathrm{d}(e^{2A}\,v) &= 0\,, &\text{(5.2.4a)} \\
e^{-3A}\mathrm{d}(e^{3A}\Omega) &= 0\,, &\text{(5.2.4b)} \\
e^{-4A}\mathrm{d}(e^{4A}\,J) &= *G\,, &\text{(5.2.4c)} \\
e^{-2A}\mathrm{d}(e^{2A}\,J\wedge J) &= -2v\wedge G\,, &\text{(5.2.4d)} \\
\mathrm{d}(J\wedge J\wedge J) &= -6v\wedge J\wedge G\,. &\text{(5.2.4e)}
\end{align}
$$

5.2.2 Conditions on the flux

Acting on (5.2.3a) with $\eta^\dagger_+ \gamma_{n_1\ldots n_{p-1}}$ and $\eta^T_+ \gamma_{n_1\ldots n_{p-1}}$ gives various constraints on the flux which we listed in appendix A.5. The most important ones of these are the three restrictions

$$
\tilde{\Sigma}_3 \lrcorner\, G = 0\,, \qquad \tilde{\Sigma}_4 \lrcorner\, G = 0\,, \qquad \Sigma_5 \lrcorner\, G = -6\,\Sigma_0\,\mathrm{d}A\,. \tag{5.2.5}
$$

Splitting the flux G into parts proportional and perpendicular to v, $G = F - v\wedge H$, and decomposing F and H under SU(3)

$$
\begin{align}
F &= A_1\, J\wedge J + A_2 \wedge J + \overline{B}\wedge\Omega + B\wedge\overline{\Omega}\,, \tag{5.2.6}\\
H &= \overline{C}_1\,\Omega + C_1\overline{\Omega} + C_2\wedge J + C_3\,,
\end{align}
$$

one finds that $B = 0$ and $C_1 = 0$.[6] Hence F is $(2,2)$ and H is $(2,1)+(1,2)$. Furthermore the exterior derivative of the warp factor is determined by A_1 and C_2

$$
\mathrm{d}A = A_1\, v - \tfrac{1}{3} C_2 \lrcorner\, J = a_1\, v + a_2\,. \tag{5.2.7}
$$

Plugging the decomposition of $\mathrm{d}v$, $\mathrm{d}J$, and $\mathrm{d}\Omega$ in terms of torsion classes (3.3.17) and (5.2.6) into (5.2.4) one can rewrite all SUSY conditions in terms of these torsion classes

$$
\begin{align}
&R = 0\,, &&V_1 = T_1 = 0\,, &&W_0 = 2\,a_2\,, &\text{(5.2.8)}\\
&E = \mathrm{Re}\,E = -3\,A_1 = -3\,a_1\,, &&W_1 = -\tfrac{4}{3}C_1 = 0 = R\,, &&V_2 = B = 0\,, \\
&W_4 = -\tfrac{1}{2}W_0 = \tfrac{1}{3}C_2\lrcorner\, J\,, &&W_2 = S = 0\,, &&T_2 = -A_2\,, \\
&2\,\mathrm{Re}\,W_5 = C_2\lrcorner\, J\,, &&2\,\mathrm{Im}\,W_5 = C_2\,, &&C_3 = -v\lrcorner *W_3\,.
\end{align}
$$

---

[5] To obtain these simple expressions one also have to make use of (5.2.3a).
[6] A_1 is a real and C_1 a complex scalar, respectively. With respect to the almost complex structure defined by J and Ω, A_2 is primitive and $(1,1)$, B is $(1,0)$, C_2 a real one-form, and C_3 is $(2,1)+(1,2)$ and primitive.

5.3 Is a BPS-like potential possible?

We have now the two basic ingredients at hand to discuss whether a BPS-like potential is possible for heterotic M-theory. On the one hand, we have rewritten the Ricci scalar in terms of exterior derivatives of v, J, and Ω. On the other hand, we know what results these derivatives have for unbroken SUSY. Thus, a BPS-like form of the potential is possible when (2.4.8) can be written as a sum of perfect squares containing the supersymmetry conditions (5.2.4). In order to check for this, we will first consider the bulk potential and turn afterwards to the boundary contributions.

5.3.1 Bulk potential

Since we know from section 5.2 that a supersymmetric vacuum must have $\mu = w_0 = 0$ we focus on these settings. This means that $\hat{R}^{(4)}$ and the both terms containing μ vanish in (2.4.9a) and we can start with

$$V_0 = -\frac{1}{2\kappa^2} \int_M \mathrm{dvol}_7 \, e^{4A} \left(R - 8\nabla^2 A - 20 \, \mathrm{d}A^2 - \frac{1}{2}|G|^2 \right). \tag{5.3.1}$$

Comparing the formula for R (5.1.4) with (5.2.4) we see that in order to possibly match the differential supersymmetry conditions we have to insert the right powers of e^A into the exterior derivatives of v, J, and Ω, respectively. This will obviously lead to terms linear in $\mathrm{d}A$. Defining

$$\begin{aligned} \mathrm{d}\tilde{v} &= e^{-2A}\mathrm{d}(e^{2A}\,v)\,, & \mathrm{d}\tilde{\Omega} &= e^{-3A}\mathrm{d}(e^{3A}\,\Omega)\,, \\ \mathrm{d}\tilde{J} &= e^{-4A}\mathrm{d}(e^{4A}\,J)\,, & \mathrm{d}\tilde{J}^2 &= e^{-2A}\mathrm{d}(e^{2A}\,J^2)\,, \end{aligned} \tag{5.3.2}$$

we find in particular that

$$\begin{aligned} \mathrm{d}\phi(\mathrm{d}v, \mathrm{d}\Omega, \mathrm{d}J) &= \mathrm{d}\phi(\mathrm{d}\tilde{v}, \mathrm{d}\tilde{\Omega}, \mathrm{d}\tilde{J}) - 6\,\mathrm{d}A \wedge v \wedge J - 3\,\mathrm{d}A \wedge \mathrm{Re}\,\Omega\,, \\ \mathrm{d}\psi(\mathrm{d}v, \mathrm{d}\Omega, \mathrm{d}J^2) &= \mathrm{d}\psi(\mathrm{d}\tilde{v}, \mathrm{d}\tilde{\Omega}, \mathrm{d}\tilde{J}^2) - \mathrm{d}A \wedge J^2 - 5\,\mathrm{d}A \wedge v \wedge \mathrm{Im}\,\Omega\,. \end{aligned} \tag{5.3.3}$$

Additional linear terms will come from the derivative piece of R in (5.1.3) after a partial integration. As we explained in section 2.1 there will be no boundary terms from the partial integration if we use the upstairs picture. We can then write[7]

---

[7]Here, we used $(\mathrm{d}J \lrcorner v) \lrcorner \mathrm{Re}\,\Omega = -\mathrm{Re}\,\mathrm{d}\Omega \lrcorner (v \wedge J)$.

$$\int_M \mathrm{dvol}_7\, e^{4A} \left\{ R - 8\nabla^2 A - 20\,\mathrm{d}A^2 \right\} \tag{5.3.4}$$

$$= \int_M \mathrm{dvol}_7\, e^{4A} \left\{ \frac{1}{2}\left|\mathrm{d}\tilde{\psi}\lrcorner\psi\right|^2 - \frac{1}{2}\left|\mathrm{d}\tilde{\phi}\right|^2 - \frac{1}{2}\left|\mathrm{d}\tilde{\psi}\right|^2 + \frac{1}{8}\left|\mathrm{d}\tilde{\phi}\lrcorner\psi\right|^2 \right.$$
$$- 18\left|\mathrm{d}A\lrcorner v\right|^2 + 3\,\mathrm{Re}\,\mathrm{d}\tilde{\Omega}\lrcorner\left(\mathrm{d}A\wedge v\wedge J\right) - 3(\mathrm{d}A\lrcorner v)\,\mathrm{d}\tilde{J}\lrcorner\mathrm{Re}\,\Omega + 6\,\mathrm{d}\tilde{v}\lrcorner(\mathrm{d}A\wedge v)$$
$$\left. + \frac{3}{2}(\mathrm{d}A\lrcorner v)\,\mathrm{Re}\left[\mathrm{d}\tilde{\Omega}\lrcorner(v\wedge\overline{\Omega})\right] - 3(\mathrm{d}A\wedge\mathrm{d}\tilde{v})\lrcorner\mathrm{Im}\,\Omega \right\}.$$

Here, $\mathrm{d}\tilde{\phi}$ and $\mathrm{d}\tilde{\psi}$ are shorthand notations for $\mathrm{d}\phi(\mathrm{d}\tilde{v}, \mathrm{d}\tilde{\Omega}, \mathrm{d}\tilde{J})$ and $\mathrm{d}\psi(\mathrm{d}\tilde{v}, \mathrm{d}\tilde{\Omega}, \mathrm{d}\tilde{J}^2)$, respectively. Clearly, all but the first four terms of this expression vanish at most linear when the conditions (5.2.4) are imposed. If it is not possible to cancel them, then a BPS-like form of V will not be available.

In order to see if such a cancellation happens, we have to include the flux in our discussion. The first four terms of (5.3.4) contain exterior derivatives $\mathrm{d}\tilde{J}$ and $\mathrm{d}\tilde{J}^2$. If SUSY is to be maintained after the compactification these should be proportional to the flux G. Inserting G will also lead to contributions that do not vanish quadratically under SUSY. Schematically these contributions will look like

$$|\mathrm{d}J\lrcorner U + V|^2 = |(\mathrm{d}J - *G)\lrcorner U + V|^2 - |*G\lrcorner U|^2 - 2(*G\lrcorner U)\lrcorner(\mathrm{d}J\lrcorner U + V), \tag{5.3.5}$$

and could eventually cancel the terms in (5.3.4). But as it turns out, a direct insertion is very cumbersome and not very enlightening.

Instead, we split the derivatives of v, J, and Ω in their parts proportional and perpendicular to v

$$\begin{aligned} \mathrm{d}\tilde{v} &= \mathrm{d}\tilde{v}_\perp + v\wedge(\mathrm{d}\tilde{v}\lrcorner v), & \mathrm{d}\tilde{\Omega} &= \mathrm{d}\tilde{\Omega}_\perp + v\wedge(\mathrm{d}\tilde{\Omega}\lrcorner v), \\ \mathrm{d}\tilde{J} &= \mathrm{d}\tilde{J}_\perp + v\wedge(\mathrm{d}\tilde{J}\lrcorner v), & \mathrm{d}\tilde{J}^2 &= \mathrm{d}\tilde{J}^2_\perp + v\wedge(\mathrm{d}\tilde{J}^2\lrcorner v). \end{aligned} \tag{5.3.6}$$

In particular, it is the fact that $\mathrm{d}\tilde{J}^2_\perp$ will vanish for supersymmetric vacua due to $\mathrm{d}\tilde{J}^2 = -4v\wedge G$ that simplifies the calculation. The square terms in (5.3.4) can be brought to the form

$$-\frac{1}{2}|\mathrm{d}\tilde{\phi}|^2 = -\frac{1}{2}|\mathrm{d}\tilde{\phi}_\perp|^2 - (\mathrm{Re}\,\mathrm{d}\tilde{\Omega}\lrcorner v)\lrcorner\left[\frac{1}{2}\mathrm{Re}\,\mathrm{d}\tilde{\Omega}\lrcorner v + (\mathrm{d}\tilde{v}\lrcorner v)\wedge J - \mathrm{d}\tilde{J}_\perp\right], \tag{5.3.7a}$$

$$-\frac{1}{2}|\mathrm{d}\tilde{\psi}|^2 = -\frac{1}{2}|\mathrm{d}\tilde{\psi}_\perp|^2 + \mathrm{Im}\mathrm{d}\tilde{\Omega}_\perp \lrcorner [J \wedge (\mathrm{d}\tilde{J}\lrcorner v)] - 3(\mathrm{d}A\lrcorner v)\,\mathrm{Im}\mathrm{d}\tilde{\Omega}_\perp \lrcorner J^2 \qquad (5.3.7\mathrm{b})$$
$$-\frac{1}{8}|\mathrm{d}\tilde{J}^2\lrcorner v|^2 - \mathrm{Re}\,\mathrm{d}\tilde{\Omega}\lrcorner (v \wedge J \wedge (\mathrm{d}\tilde{v}\lrcorner v))\,,$$

$$\frac{1}{2}|\mathrm{d}\tilde{\psi}\lrcorner\psi|^2 = \frac{1}{2}|\mathrm{d}\tilde{\psi}_\perp\lrcorner\psi|^2 + \frac{1}{8}\left[\mathrm{d}\tilde{J}^2\lrcorner(v\wedge J^2)\right]\lrcorner\left[\frac{1}{4}\mathrm{d}\tilde{J}^2\lrcorner(v\wedge J^2) - (\mathrm{Im}\mathrm{d}\tilde{\Omega}_\perp\lrcorner J^2)\right] \qquad (5.3.7\mathrm{c})$$
$$+\left[\mathrm{Re}\,\mathrm{d}\tilde{\Omega}\lrcorner(v\wedge J)\right]\lrcorner\left[\frac{1}{2}\mathrm{Re}\,\mathrm{d}\tilde{\Omega}\lrcorner(v\wedge J) - \mathrm{d}\tilde{v}_\perp\lrcorner\mathrm{Im}\Omega - \frac{1}{4}\mathrm{d}\tilde{J}_\perp^2\lrcorner J^2 + \frac{1}{2}\mathrm{Re}(\mathrm{d}\tilde{\Omega}_\perp\lrcorner\bar{\Omega}) + 2\mathrm{d}\tilde{v}\lrcorner v\right]\,,$$

$$\frac{1}{8}|\mathrm{d}\tilde{\phi}\lrcorner\psi|^2 = \frac{1}{8}|\mathrm{d}\tilde{\phi}_\perp\lrcorner\psi|^2 + \frac{1}{8}\mathrm{Im}[\mathrm{d}\tilde{\Omega}\lrcorner(v\wedge\bar{\Omega})] \times \left[\frac{1}{4}\mathrm{Im}[\mathrm{d}\tilde{\Omega}\lrcorner(v\wedge\bar{\Omega})] - 2\mathrm{d}\tilde{v}_\perp\lrcorner J - \mathrm{Re}\,\mathrm{d}\tilde{\Omega}_\perp\lrcorner J^2\right]. \qquad (5.3.7\mathrm{d})$$

Note that in each expression there is one term including $\mathrm{d}\tilde{\phi}_\perp = \mathrm{d}\phi(\mathrm{d}\tilde{v}, \mathrm{d}\tilde{\Omega}_\perp, \mathrm{d}\tilde{J}_\perp)$ or $\mathrm{d}\tilde{\psi}_\perp = \mathrm{d}\psi(\mathrm{d}\tilde{v}, \mathrm{d}\tilde{\Omega}_\perp, \mathrm{d}\tilde{J}_\perp^2)$. We find that in the combination of these

$$\frac{1}{2}\left|\mathrm{d}\tilde{\psi}_\perp\lrcorner\psi\right|^2 + \frac{1}{8}\left|\mathrm{d}\tilde{\phi}_\perp\lrcorner\psi\right|^2 - \frac{1}{2}\left|\mathrm{d}\tilde{\phi}_\perp\right|^2 - \frac{1}{2}\left|\mathrm{d}\tilde{\psi}_\perp\right|^2 = \qquad (5.3.8)$$
$$-\frac{1}{2}|\mathrm{d}\tilde{J}_\perp|^2 - \frac{1}{2}|\mathrm{d}\tilde{\Omega}_\perp|^2 - \frac{1}{8}|\mathrm{d}\tilde{J}_\perp^2|^2 + \frac{1}{2}\left|\frac{1}{4}\mathrm{d}\tilde{J}_\perp^2\lrcorner J^2 - \frac{1}{2}\mathrm{Re}(\mathrm{d}\tilde{\Omega}_\perp\lrcorner\bar{\Omega}) - \mathrm{d}\tilde{v}\lrcorner v\right|^2 + \frac{1}{8}\left|\mathrm{d}\tilde{\Omega}_\perp\lrcorner J^2\right|^2$$
$$-\frac{1}{2}\left|\mathrm{d}\tilde{v}\lrcorner v\right|^2 - \frac{1}{2}\left|\mathrm{d}\tilde{v}_\perp\right|^2 - 2\left(\mathrm{d}\tilde{v}_\perp\lrcorner\mathrm{Im}\Omega\right)\lrcorner(\mathrm{d}\tilde{v}\lrcorner v) + \frac{1}{6}(\mathrm{Re}\,\mathrm{d}\tilde{\Omega}_\perp\wedge\mathrm{d}\tilde{v}_\perp)\lrcorner J^3 - 6\mathrm{d}\tilde{v}\lrcorner(\mathrm{d}A\wedge v)\,,$$

only the last term and $|\mathrm{d}\tilde{J}_\perp|^2$ do not vanish quadratically when supersymmetry is imposed. Note that in order to obtain this expression we used the identities

$$\mathrm{Im}\,\mathrm{d}\tilde{\Omega}_\perp\lrcorner\left[(\mathrm{d}\tilde{v}_\perp\lrcorner\mathrm{Im}\Omega)\wedge\mathrm{Im}\Omega\right] + \frac{1}{2}(\mathrm{d}\tilde{v}_\perp\lrcorner J)(\mathrm{Re}\,\mathrm{d}\tilde{\Omega}_\perp\lrcorner J^2) - \mathrm{Re}\,\mathrm{d}\tilde{\Omega}_\perp\lrcorner(\mathrm{d}\tilde{v}_\perp\wedge J) = \qquad (5.3.9)$$
$$= \frac{1}{6}(\mathrm{Re}\,\mathrm{d}\tilde{\Omega}_\perp\wedge\mathrm{d}\tilde{v}_\perp)\lrcorner J^3\,,$$
$$|\mathrm{d}v\wedge\mathrm{Im}\Omega|^2 - |\mathrm{d}v\lrcorner\mathrm{Im}\Omega|^2 = 2|\mathrm{d}v\lrcorner v|^2\,,\quad |\mathrm{d}v\wedge J|^2 - |\mathrm{d}v\lrcorner J|^2 = |\mathrm{d}v_\perp|^2 + 2|\mathrm{d}v\lrcorner v|^2\,.$$

Since in the end we will integrate over the whole expression, we can even get a further simplification using partial integration and the fact that $2(\mathrm{d}\tilde{v}_\perp\lrcorner\mathrm{Im}\,\Omega)\lrcorner(\mathrm{d}\tilde{v}\lrcorner v) = (\mathrm{d}\tilde{v}\wedge\mathrm{d}\tilde{v})\lrcorner(v\wedge\mathrm{Im}\,\Omega)$

$$\int_M e^{4A}\left\{\frac{1}{6}(\mathrm{Re}\,\mathrm{d}\tilde{\Omega}_\perp\wedge\mathrm{d}\tilde{v}_\perp)\lrcorner J^3 - (\mathrm{d}\tilde{v}\wedge\mathrm{d}\tilde{v})\lrcorner(v\wedge\mathrm{Im}\Omega)\right\} = 3\int_M e^{4A}\,(\mathrm{d}A\wedge\mathrm{d}\tilde{v})\lrcorner\mathrm{Im}\Omega\,. \qquad (5.3.10)$$

This will cancel exactly against the last term appearing in (5.3.4). We thus conclude that we can neglect all terms including $\mathrm{d}\tilde{v}_\perp$ in (5.3.8) except for $-\frac{1}{2}|\mathrm{d}\tilde{v}_\perp|^2$, as long as we also neglect the term

$-3(\mathrm{d}A \wedge \mathrm{d}\tilde{v}) \lrcorner \mathrm{Im}\Omega$ from (5.3.4). We also see that the last term of (5.3.8) will cancel against a term in (5.3.4).

Examining the rest of (5.3.7a) – (5.3.7c), we find only six more terms that do not vanish quadratically under supersymmetry

$$\mathrm{Re}\,\mathrm{d}\tilde{\Omega}\lrcorner (v \wedge \mathrm{d}\tilde{J}_\perp)\,, \qquad \left[J \wedge (\mathrm{d}\tilde{J}\lrcorner v)\right]\lrcorner \mathrm{Im}\mathrm{d}\tilde{\Omega}_\perp\,, \qquad -\frac{1}{8}\left|\mathrm{d}\tilde{J}^2\lrcorner v\right|^2, \qquad (5.3.11)$$
$$\frac{1}{32}\left|\mathrm{d}\tilde{J}^2\lrcorner (v \wedge J^2)\right|^2, \qquad -\frac{1}{8}\left[\mathrm{d}\tilde{J}^2\lrcorner (v \wedge J^2)\right]\lrcorner \left(\mathrm{Im}\tilde{\Omega}_\perp\lrcorner J^2\right), \qquad -3(\mathrm{d}A\lrcorner v)\mathrm{Im}\mathrm{d}\tilde{\Omega}\lrcorner J^2\,.$$

This means that due to the split (5.3.6) we have reduced the number of squares that do not vanish under SUSY at all, and which thus should be combined with G flux, to three. To check whether from these terms can come contributions that cancel the linearly vanishing expressions, or if some of them cancel themselves, we will in the end expand all expressions in terms of the SU(3) torsion classes (3.3.17) and use the SUSY conditions in the form (5.2.8).

But before we do so, it is important to note that we have not yet used the Bianchi identity of the four-form flux G. In [197] and in chapter 4 the use of the Bianchi identity was crucial in order to obtain a BPS-like potential. We follow these references and implement the Bianchi identity by a partial integration

$$\int_M \mathrm{dvol}_7\, e^{4A}\left\{-\frac{1}{2}|\mathrm{d}\tilde{J}_\perp|^2 - \frac{1}{2}|G|^2\right\} = \int_M \mathrm{dvol}_7\, e^{4A}\left\{-\frac{1}{2}|\mathrm{d}\tilde{J}|^2 - \frac{1}{2}|G|^2 + \frac{1}{2}|\mathrm{d}\tilde{J}\lrcorner v|^2\right\} =$$
$$\int_M \mathrm{dvol}_7\, e^{4A}\left\{-\frac{1}{2}|\mathrm{d}\tilde{J} - *G|^2 + \frac{1}{2}\mathrm{d}G\lrcorner(v \wedge J^2) + \frac{1}{2}|\mathrm{d}\tilde{J}\lrcorner v|^2\right\} = \qquad (5.3.12)$$
$$\int_M \mathrm{dvol}_7\, e^{4A}\left\{-\frac{1}{2}|\mathrm{d}\tilde{J} - *G|^2 + \frac{1}{2}|\mathrm{d}\tilde{J}\lrcorner v|^2\right\} - \frac{1}{4\pi}\left(\frac{\kappa}{4\pi}\right)^{2/3}\sum_{p=1,2}\int_{B_{6,p}} e^{4A}\, J \wedge K^{(p)},$$

where we used (2.1.8) in the last step. We conclude that we have to include $\frac{1}{2}|\mathrm{d}\tilde{J}\lrcorner v|^2$ into our analysis of the bulk action in order to take the Bianchi identity of G into account. Furthermore, we get an additional contribution to the boundary action, due to the δ-terms of the BI (2.1.8).

All things considered, in order to obtain a BPS-like form of the potential the sum of $\frac{1}{2}|\mathrm{d}\tilde{J}\lrcorner v|^2$, the remaining linear terms of (5.3.4), and the six terms of (5.3.11)

$$L = \frac{1}{2}|\mathrm{d}\tilde{J}\lrcorner v|^2 - \frac{1}{8}|\mathrm{d}\tilde{J}^2\lrcorner v|^2 + \frac{1}{32}\left|\mathrm{d}\tilde{J}^2\lrcorner(v \wedge J^2)\right|^2 - \frac{1}{8}\left[\mathrm{d}\tilde{J}^2\lrcorner(v \wedge J^2)\right]\lrcorner(\mathrm{Im}\tilde{\Omega}_\perp\lrcorner J^2)$$
$$- 18|\mathrm{d}A\lrcorner v|^2 + \left[J \wedge (\mathrm{d}\tilde{J}\lrcorner v)\right]\lrcorner \mathrm{Im}\mathrm{d}\tilde{\Omega}_\perp + \mathrm{Re}\,\mathrm{d}\tilde{\Omega}\lrcorner(v \wedge \mathrm{d}\tilde{J}_\perp) + 3\,\mathrm{Re}\,\mathrm{d}\tilde{\Omega}\lrcorner(\mathrm{d}A \wedge v \wedge J)$$
$$- 3(\mathrm{d}A\lrcorner v)\,\mathrm{d}\tilde{J}\lrcorner\mathrm{Re}\Omega - 3(\mathrm{d}A\lrcorner v)\,\mathrm{Im}\,\mathrm{d}\tilde{\Omega}\lrcorner J^2 + \frac{3}{2}(\mathrm{d}A\lrcorner v)\,\mathrm{Re}\left[\mathrm{d}\tilde{\Omega}\lrcorner(v \wedge \overline{\Omega})\right] \qquad (5.3.13)$$

has to vanish quadratically for a supersymmetric setting. Inserting the expansion (3.3.17), and reordering terms we get

$$\begin{aligned}L &= \frac{3}{2}\left|(\mathrm{d}\tilde{J}\lrcorner(v\wedge J)-6\,\mathrm{d}A\lrcorner v\right|^2 + (\mathrm{Re}\mathrm{d}\tilde{\Omega}\lrcorner v)\lrcorner[\mathrm{d}\tilde{J}-3\,\mathrm{d}A\wedge J]\\ &\quad + \mathrm{Im}\mathrm{d}\tilde{\Omega}\lrcorner(J\wedge(\mathrm{d}\tilde{J}\lrcorner v)) - 24\,\mathrm{Re}E\,\mathrm{Im}\,W_1 - 90\,(\mathrm{d}A\lrcorner v)\mathrm{Im}\,W_1 \quad (5.3.14)\\ &= 6\,|\mathrm{Re}E+3\,\mathrm{d}A\lrcorner v|^2 - 10\,\mathrm{Im}\,W_1(\mathrm{Re}E+3\,\mathrm{d}A\lrcorner v) - 8\,\mathrm{Re}\,V_2\lrcorner(2\,\mathrm{Re}\,W_5+W_4+4\,\mathrm{d}A)\\ &\quad + 6\,\mathrm{Im}\,E\,\mathrm{Re}\,W_1 + 6\,(\mathrm{d}A\lrcorner v)\mathrm{Im}\,W_1 + T_2\lrcorner\mathrm{Im}\,W_2 + \mathrm{Re}\,S\lrcorner W_3\ . \quad (5.3.15)\end{aligned}$$

Using the relations (5.2.8) we see that all except of the last three terms of (5.3.15) will indeed go to zero quadratically under SUSY. The last three terms vanishes linearly, since $\mathrm{d}A\lrcorner v$, T_2, and W_3 are non-zero generically. However, we can rewrite $(\mathrm{Re}E\,\mathrm{Im}W_1)$ using partial integration and the fact that $\mathrm{Im}\,\mathrm{d}W_1\lrcorner v = 0$ under SUSY[8]

$$\int_M \mathrm{dvol}_7\, e^{4A}\,\mathrm{Re}E\,\mathrm{Im}W_1 = -2\int_M \mathrm{dvol}_7\, e^{4A}\,(\mathrm{d}A\lrcorner v)\mathrm{Im}W_1\ . \quad (5.3.16)$$

Another partial integration shows that $(\mathrm{d}A\lrcorner v)\mathrm{Im}W_1$ will vanish quadratically under the integral, since

$$\int_M \mathrm{dvol}_7\, e^{4A}\,(\mathrm{d}A\lrcorner v)\mathrm{Im}W_1 = -\frac{1}{4}\int_M \mathrm{dvol}_7\, e^{4A}\bigl[(\partial^m v_m)\,\mathrm{Im}W_1+\mathrm{Im}\,\mathrm{d}W_1\lrcorner v\bigr]\ . \quad (5.3.17)$$

The second term on the right hand side gives zero, and therefore one obtains

$$\int_M \mathrm{dvol}_7\, e^{4A}\bigl[(\mathrm{d}A\lrcorner v)+\frac{1}{4}(\partial^m v_m)\bigr]\mathrm{Im}W_1 = 0\ . \quad (5.3.18)$$

$\mathrm{Im}W_1$ is an arbitrary function (not depending on the direction of v) and can be viewed as a test function. Thus $e^{4A}\bigl[(\mathrm{d}A\lrcorner v)+\frac{1}{4}(\partial^m v_m)\bigr]$ integrates to zero. But under SUSY (5.2.3b) gives $\partial^m v_m = \nabla^m v_m = 7(\mathrm{d}A\lrcorner v)$ which means that $e^{4A}(\mathrm{d}A\lrcorner v)$ will also integrate to zero when SUSY is imposed. Since $\mathrm{Im}W_1$ is zero in this case, too, we see that $\int \mathrm{dvol}_7\, e^{4A}\,(\mathrm{d}A\lrcorner v)\mathrm{Im}W_1$ will vanish quadratically in a supersymmetric setting.

Unfortunately, we cannot see how one could argue in a similar way for the last two terms of (5.3.15). Thus, we conclude with the surprising result that M-theory compactified on a general seven-dimensional SU(3) structure manifold does not admit a BPS-like scalar potential, since in general the torsion classes T_2 and W_3 do not vanish.

---

[8]Here, we used $\mathrm{Re}E\,\mathrm{dvol}_7 = \frac{1}{2}\mathrm{d}J\lrcorner(v\wedge J)\,\mathrm{dvol}_7 = \frac{1}{4}\mathrm{d}J\wedge J\wedge J = \frac{1}{12}\mathrm{d}J^3$.

Gathering all terms together the bulk potential reads

$$V_0 = \frac{1}{4\kappa^2}\int_M \mathrm{dvol}_7\, e^{4A}\Big\{|\mathrm{d}\tilde{J} - *G|^2 + |\mathrm{d}\tilde{\Omega}_\perp|^2 + \frac{1}{4}|\mathrm{d}\tilde{J}_\perp^2|^2 - \frac{1}{4}|\mathrm{d}\tilde{\Omega}_\perp \lrcorner J^2|^2 \tag{5.3.19}$$

$$-\Big|\frac{1}{4}\mathrm{d}\tilde{J}_\perp^2 \lrcorner J^2 - \frac{1}{2}\mathrm{Re}(\mathrm{d}\tilde{\Omega}_\perp \lrcorner \bar{\Omega}) - \mathrm{d}\tilde{v}\lrcorner v - \mathrm{Re}\,\mathrm{d}\tilde{\Omega}\lrcorner(v\wedge J)\Big|^2 + |\mathrm{d}\tilde{v}\lrcorner v|^2 + |\mathrm{d}\tilde{v}_\perp|^2$$

$$+\big|\mathrm{Re}\,\mathrm{d}\tilde{\Omega}\lrcorner v\big|^2 + \frac{1}{4}\mathrm{Im}[\mathrm{d}\tilde{\Omega}\lrcorner(v\wedge \bar{\Omega})]\times\Big[\frac{1}{4}\mathrm{Im}[\mathrm{d}\tilde{\Omega}\lrcorner(v\wedge \bar{\Omega})] - 2\,\mathrm{d}\tilde{v}_\perp \lrcorner J - \mathrm{Re}\,\mathrm{d}\tilde{\Omega}_\perp \lrcorner J^2\Big]$$

$$+2\,\mathrm{Re}\,\mathrm{d}\tilde{\Omega}\lrcorner\big(v\wedge J\wedge(\mathrm{d}\tilde{v}_\perp\lrcorner\mathrm{Im}\,\Omega)\big) - 3\big|\mathrm{d}\tilde{J}\lrcorner(v\wedge J) - 6\,\mathrm{d}A\lrcorner v\big|^2 + 7\,(\mathrm{d}A\lrcorner v)\mathrm{Im}\,\mathrm{d}\tilde{\Omega}_\perp \lrcorner J^2$$

$$-2\,(\mathrm{Re}\,\mathrm{d}\tilde{\Omega}\lrcorner v)\lrcorner\big[\mathrm{d}\tilde{J}_\perp - 3\,\mathrm{d}A\wedge J\big] - 2\,\mathrm{Im}\,\mathrm{d}\tilde{\Omega}_\perp\lrcorner(J\wedge(\mathrm{d}\tilde{J}\lrcorner v))\Big\}\,.$$

One should notice here that it are the last two terms that spoil the BPS-like form

$$V_{\text{no-BPS}} = -\frac{1}{2\kappa^2}\int_M \mathrm{dvol}_7\, e^{4A}\Big\{(\mathrm{Re}\,\mathrm{d}\tilde{\Omega}\lrcorner v)\lrcorner[\mathrm{d}\tilde{J}_\perp - 3\,\mathrm{d}A\wedge J] + \mathrm{Im}\,\mathrm{d}\tilde{\Omega}_\perp\lrcorner(J\wedge(\mathrm{d}\tilde{J}\lrcorner v))\Big\}$$

$$= -\frac{1}{2\kappa^2}\int_M \mathrm{dvol}_7\, e^{4A}\Big\{24\,\mathrm{Im}\,W_1(\mathrm{d}A\lrcorner v) + \mathrm{Re}\,S\lrcorner W_3 + \mathrm{Im}\,W_2\lrcorner T_2 + 6\,\mathrm{Im}\,E\,\mathrm{Re}\,W_1$$

$$+ 14\,\mathrm{Im}\,W_1(\mathrm{Re}\,E + 3\,\mathrm{d}A\lrcorner v) - 8\,\mathrm{Re}\,V_2\lrcorner(2\,\mathrm{Re}\,W_5 + W_4 + 4\mathrm{d}A)\Big\}\,. \tag{5.3.20}$$

However, for S and T_2 identically vanishing, also this part of the potential reduces to a BPS-like form

$$V_{\text{no-BPS}} \stackrel{S=0}{\underset{T_2=0}{=}} -\frac{1}{2\kappa^2}\int_M \mathrm{dvol}_7\, e^{4A}\Big\{\frac{1}{2}\big[\mathrm{d}\tilde{J}_\perp\lrcorner J - 6\,a_2 - \frac{1}{2}\mathrm{Re}(\mathrm{d}\tilde{\Omega}_\perp\lrcorner\bar{\Omega})\big]\lrcorner[\mathrm{Re}\tilde{\Omega}\lrcorner(v\wedge J)]$$

$$+ \frac{1}{16}\mathrm{Im}[\mathrm{d}\tilde{\Omega}\lrcorner(v\wedge\bar{\Omega})]\,\mathrm{Re}\,\mathrm{d}\tilde{\Omega}_\perp\lrcorner J^2 + \frac{19}{6}(\mathrm{d}A\lrcorner v)\mathrm{Im}\,\mathrm{d}\tilde{\Omega}_\perp\lrcorner J^2\Big\}\,. \tag{5.3.21}$$

So we wee that it is in general not possible to bring the bulk part of the potential to a BPS-like form. But by setting the terms containing T_2 and W_3 to zero, such a form can be reached.

5.3.2 Boundary potential

The boundary potential receives contributions from two sources. Besides (2.4.9b) one also has to include the piece obtained by integration over the Bianchi identity in (5.3.12). Before combining these two expressions, one has to make sure that both are given in terms of the same metric g'_{10} that we introduced in section 2.4. However, going from g_{10} to g'_{10} does not lead to new contributions from $K^{(p)}$, and thus one simply finds

$$V_b = \frac{1}{8\pi\kappa^2}\left(\frac{\kappa}{4\pi}\right)^{2/3} \sum_{p=1,2} \int_{B_{6,p}} \mathrm{dvol}'_{6,p}\, e^{4A'-2\phi} \Bigg\{ -\frac{1}{24}\left|e^{-2A'}\hat{R}^{(4)} - 12|\mathrm{d}A'|^2\right|^2 \tag{5.3.22}$$

$$+ \left(\mathrm{Tr}|\mathcal{F}_p^{(6)}\lrcorner J|^2 + 2\,\mathrm{Tr}|(\mathcal{F}_p^{(6)})^{2,0}|^2\right) - \frac{1}{2}\left(\mathrm{Tr}|R_{p+}^{(6)'}\lrcorner J|^2 + 2\,\mathrm{Tr}|(R_{p+}^{(6)'})^{2,0}|^2\right)$$

$$- 4\,e^{-2A'}\left(\nabla_i\nabla_j e^{A'}\right)(\nabla^i\nabla^j e^{A'}) - 2\left|\mathrm{d}A'\lrcorner H\right|^2 \Bigg\}.$$

Since for a supersymmetric vacuum we have to restrict to Minkowski space, $\hat{R}^{(4)}$ will vanish. Also the terms containing $\mathcal{F}_p$ will vanish by the SUSY conditions (5.2.2). Let us consider next the $\mathrm{d}A'$ terms. From (5.2.4a), (5.2.7), and the definition $A' = A + \frac{1}{2}\sigma$ given in section 2.4.1 we know that

$$-2\,\mathrm{d}A' + \mathrm{d}\sigma_\perp + \mathrm{d}v\lrcorner v = 0, \tag{5.3.23}$$

where $\mathrm{d}\sigma_\perp$ denotes as usual the part of $\mathrm{d}\sigma$ that is perpendicular to v. Thus, in order to obtain $\mathrm{d}A' = 0$ for a SUSY vacuum, the identity $\mathrm{d}\sigma_\perp = -W_0$ should hold. Since we did not specify σ yet, we choose it in such a way that the above equation is satisfied. In section 5.4.3 we will see a justification for this choice.

The last terms to consider are the $R_{p+}^{(6)'}$ terms. These vanish if $R_{p+}^{(6)'}$ is $(1,1)$ and primitive. For the heterotic string this can be shown by using the integrability condition

$$\left[\nabla_i^{\mathrm{het}}, \nabla_j^{\mathrm{het}}\right]\eta_{\mathrm{het}} = \frac{1}{4} R_{klij}\gamma^{kl}\,\eta_{\mathrm{het}}, \tag{5.3.24}$$

where 'het' denotes that the various objects belong to ten-dimensional heterotic supergravity (see e.g. [137, 221] and chapter 4). We thus trace back the problem to the ten-dimensional setting.

In order to do so, one has to determine how the seven-dimensional covariant derivative ∇_m, which appears in (5.2.3b), is related to its six-dimensional counterpart at the boundary. This is an quite easy task in type IIA supergravity, where the geometry in ten dimensions can be chosen to be independent of the extra dimension (see e.g. [10]). In Horava-Witten theory this gets changed by the non-vanishing of the four-form flux Bianchi identity. However, the modifications appear only at order $\kappa^{2/3}$. Since the boundary terms are already of order $\kappa^{2/3}$, one can consistently neglect the corrections and work in the type IIA setting

$$\mathrm{d}s_7^2 = e^{-\sigma}(\mathrm{d}s_6')^2 + e^{2\sigma}\mathrm{d}x^{11} \tag{5.3.25}$$

with σ and g'_6 independent of x^{11}. A calculation along the lines of [10] shows then that

$$\nabla^{(6)'}_{-i}\left(e^{\frac{\sigma}{4}}\eta_+\right) = \nabla^{(6)'}_{i}\left(e^{\frac{\sigma}{4}}\eta_+\right) - \frac{1}{8}G_{11\,ijk}\gamma^{jk}\left(e^{\frac{\sigma}{4}}\eta_+\right) = \mathcal{O}(\kappa^{2/3})\,. \qquad (5.3.26)$$

Then, on each boundary

$$\left[\nabla^{(6)'}_{-i},\nabla^{(6)'}_{-j}\right]\left(e^{\frac{\sigma}{4}}\eta_+\right) = e^{\frac{\sigma}{4}}\left[\nabla^{(6)'}_{-i},\nabla^{(6)'}_{-j}\right]\eta_+ = \frac{1}{4}e^{\frac{\sigma}{4}}R^{(6)'}_{-klij}\gamma^{kl}\eta_+ = \mathcal{O}(\kappa^{2/3})\,, \qquad (5.3.27)$$

which is precisely the condition one obtains for the heterotic string. From this it follows that $R^{(6)'}_{p+}$ is $(1,1)$ and primitive up to $\mathcal{O}(\kappa^{2/3})$, which is sufficient for our analysis as the boundary potential is already of order $\kappa^{2/3}$.

We conclude that the boundary potential can be rewritten in a BPS-like form, although this is not possible for the bulk potential. The fact that a BPS-like form of the potential is not available in general does of course not mean that our ansatz is inconsistent. It merely states that in addition to the Bianchi identity and the SUSY conditions the ansatz has to satisfy further restrictions that come from the variation of (5.3.20) in order to ensure the equations of motion. On the other hand, (5.3.21) tells us how to restrict our compactification ansatz if we wish to get a BPS-like scalar potential. For example if one chooses a manifold for which T_2 and $\text{Re}\,S$ vanish identically (and not only if SUSY is imposed) then the whole action can be written in terms of squares of the supersymmetry conditions. However, it would be nice to see whether our findings give the correct results when restricted to well known geometrical settings. This will be discussed in the next section.

5.4 Limiting cases

In order to strengthen our results, we will show that they reduce correctly in the three cases of G_2 holonomy, six-dimensional SU(3) holonomy, and the heterotic limit.

5.4.1 G_2 holonomy

It is well known that compactifications on manifolds with G_2 holonomy do not allow four-form flux G. Hence, they are not viable for Horava-Witten theory where G is necessarily non-zero. Nevertheless, one can ask whether our formulas behave in the right way in this limit, although we know that they will not give a solution for heterotic M-theory. In particular, we expect the curvature scalar R to be zero for a G_2 holonomy manifold M. Furthermore, once the SUSY conditions are satisfied also the G flux should be set to zero and the warp factor A should be constant.

G$_2$ holonomy is specified by the conditions

$$d\phi = 0, \quad \text{and} \quad d\psi = 0. \tag{5.4.1}$$

Applying these conditions to the decomposition (3.3.16) of ϕ and ψ in SU(3) structure forms one finds that for G$_2$ holonomy manifolds

$$\begin{aligned}
&\text{Re} W_1 = \frac{2}{3} \text{Im} E = -R, \quad \text{Im} W_1 = \frac{2}{3} \text{Re} E, \quad T_1 = -\text{Re} W_2, \quad \tilde{T}_1 = \text{Im} W_2, \\
&\text{Re} V_1 = \frac{1}{2} \text{Im} W_5, \quad \text{Im} V_1 = -\frac{1}{2} \text{Re} W_5, \quad \text{Re} S = W_3, \quad J \lrcorner W_0 = 2 \text{Im} V_2, \\
&W_0 = W_4 + 4 \text{Re} V_2 = \text{Re} W_5 + 2 \text{Re} V_2.
\end{aligned} \tag{5.4.2}$$

These conditions do clearly not imply that all SU(3) torsion classes vanish. This means that although it is clear from (5.4.1) and (5.1.3) that $R = 0$, it is a non-trivial consistency check for our equations that the scalar curvature also vanishes in (5.3.4) for a G$_2$ manifold.

In order to test also the equations (5.3.7) we split $d\phi$ and $d\psi$ into parts proportional and perpendicular to v and find the four conditions

$$\begin{aligned}
\text{Re} \, d\tilde{\Omega}_\perp &= 3 a_2 \wedge \text{Re} \, \Omega, \\
\frac{1}{2} d\tilde{J}_\perp^2 &= a_2 \wedge J^2 - d\tilde{v}_\perp \wedge \text{Im} \, \Omega, \\
\text{Im} \, d\tilde{\Omega}_\perp &= \frac{1}{2} d\tilde{J}^2 \lrcorner v + (5 a_2 + d\tilde{v} \lrcorner v) \wedge \text{Im} \, \Omega - (dA \lrcorner v) J^2, \\
d\tilde{J}_\perp &= (6 \, dA + d\tilde{v} \lrcorner v) \wedge J + \text{Re} \, d\tilde{\Omega} \lrcorner v - 3 (dA \lrcorner v) \text{Re} \, \Omega.
\end{aligned} \tag{5.4.3}$$

Plugging these into (5.3.7), one can express the first line of (5.3.4) solely in terms of $dA \lrcorner v$

$$\frac{1}{2} |d\tilde{\psi} \lrcorner \psi|^2 - \frac{1}{2} |d\tilde{\phi}|^2 - \frac{1}{2} |d\tilde{\psi}|^2 + \frac{1}{8} |d\tilde{\phi} \lrcorner \psi|^2 = -6 \, dA \lrcorner v, \tag{5.4.4}$$

while the rest of (5.3.4) reduces to $12|a_2|^2 + 18|dA \lrcorner v|^2$. We thus see that

$$\int_M \text{dvol}_7 \, e^{4A} \{R - 8\nabla^2 A - 20 \, dA^2\} = 12 \int_M \text{dvol}_7 \, e^{4A} |dA|^2, \tag{5.4.5}$$

which is just what one gets by setting $R = 0$ and integrating by parts. This means that our formulas give indeed the right results in the G$_2$ holonomy limit.

Considering the SUSY conditions, one immediately sees that the combination of the conditions

(5.2.8) and (5.4.2) leads necessarily to the vanishing of all torsion classes and all flux components. Since this is what is expected for a G_2 holonomy compactification, we conclude that also here our formulas provide the correct answer.

5.4.2 SU(3) holonomy

Next we consider manifolds M that obey

$$\mathrm{d}J_\perp = 0 \quad \text{and} \quad \mathrm{d}\Omega_\perp = 0. \tag{5.4.6}$$

This means that locally M splits into a six-dimensional manifold of SU(3) holonomy (i.e. a Calabi-Yau three-fold) times a line in the direction of v. Globally however, there can still be dependencies of J and Ω on v, and hence $\mathrm{d}J \neq 0 \neq \mathrm{d}\Omega$. In terms of torsion classes this can be achieved by setting $W_i = 0$ for $i = 1, \ldots, 5$. Since this will cancel all terms in (5.3.15) that do not vanish quadratically when SUSY is imposed, for this case a BPS-like form of the bulk potential is possible.

But before we come to the potential let us again check whether our formulas are consistent. We have now

$$\mathrm{d}\phi_{CY} = \mathrm{d}v \wedge J + v \wedge (\mathrm{Re}\,\mathrm{d}\Omega \lrcorner v), \quad \text{and} \quad \mathrm{d}\psi_{CY} = \frac{1}{2} v \wedge (\mathrm{d}J^2 \lrcorner v) + \mathrm{d}v \wedge \mathrm{Im}\,\Omega. \tag{5.4.7}$$

Plugging these directly into (5.1.3) we find

$$\int_M \mathrm{dvol}_7 \, e^{4A} R = \int_M \mathrm{dvol}_7 \, e^{4A} \left\{ \frac{1}{32} \left| \mathrm{d}J^2 \lrcorner (v \wedge J^2) \right|^2 - \frac{1}{8} \left| \mathrm{d}J^2 \lrcorner v \right|^2 - \frac{1}{2} \left| \mathrm{Re}\,\mathrm{d}\Omega \lrcorner v \right|^2 \right. \tag{5.4.8}$$

$$- \frac{1}{2} \left| \mathrm{d}v_\perp \right|^2 + \left(\mathrm{Re}\,\mathrm{d}\Omega \lrcorner (v \wedge J) \right) \lrcorner \left[\frac{1}{2} \mathrm{Re}\,\mathrm{d}\Omega \lrcorner (v \wedge J) - \mathrm{Im}\,\Omega \lrcorner \mathrm{d}v_\perp \right] - 2\,\mathrm{Im}\,\Omega \lrcorner \left((\mathrm{d}v \lrcorner v) \wedge \mathrm{d}v_\perp \right)$$

$$+ \frac{1}{8} \mathrm{Im}[\mathrm{d}\Omega \lrcorner (v \wedge \bar{\Omega})] \times \left[\frac{1}{4} \mathrm{Im}[\mathrm{d}\Omega \lrcorner (v \wedge \bar{\Omega})] - 2\mathrm{d}\tilde{v}_\perp \lrcorner J \right] - 4\,\mathrm{Re}\,\mathrm{d}\Omega \lrcorner (v \wedge a_2 \wedge J)$$

$$\left. + 4\,\mathrm{Im}\,\Omega \lrcorner (a_2 \wedge \mathrm{d}v_\perp) + 8\,\mathrm{d}v \lrcorner (a_2 \wedge v) + (\mathrm{d}A \lrcorner v) \left(\mathrm{d}J^2 \lrcorner (v \wedge J^2) \right) \right\}.$$

On the other hand, working with (5.3.4), (5.3.7), and (5.3.8) we get

$$\frac{1}{2} \left| \mathrm{d}\tilde{\psi} \lrcorner \psi \right|^2 + \frac{1}{8} \left| \mathrm{d}\tilde{\phi} \lrcorner \psi \right|^2 - \frac{1}{2} \left| \mathrm{d}\tilde{\phi} \right|^2 - \frac{1}{2} \left| \mathrm{d}\tilde{\psi} \right|^2 - 12 \left| \mathrm{d}A \right|^2 = \frac{1}{32} \left| \mathrm{d}J^2 \lrcorner (v \wedge J^2) \right|^2 \tag{5.4.9}$$

$$- \frac{1}{2} \left| \mathrm{Re}\,\mathrm{d}\Omega \lrcorner v \right|^2 - \frac{1}{8} \left| \mathrm{d}J^2 \lrcorner v \right|^2 - \frac{1}{2} \left| \mathrm{d}v_\perp \right|^2 + 2\,\mathrm{d}v \lrcorner (a_2 \wedge v) - 2\,\mathrm{Im}\,\Omega \lrcorner \left((\mathrm{d}v \lrcorner v) \wedge \mathrm{d}v_\perp \right)$$

$$+ \frac{1}{8} \mathrm{Im}[\mathrm{d}\Omega \lrcorner (v \wedge \bar{\Omega})] \times \left[\frac{1}{4} \mathrm{Im}[\mathrm{d}\Omega \lrcorner (v \wedge \bar{\Omega})] - 2\mathrm{d}\tilde{v}_\perp \lrcorner J \right] - \mathrm{Re}\,\mathrm{d}\Omega \lrcorner (v \wedge a_2 \wedge J) - 12 \left| a_2 \right|^2$$

$$+ \left(\mathrm{Re}\,\mathrm{d}\Omega \lrcorner (v \wedge J) \right) \lrcorner \left[\frac{1}{2} \mathrm{Re}\,\mathrm{d}\Omega \lrcorner (v \wedge J) - \mathrm{Im}\,\Omega \lrcorner \mathrm{d}v_\perp \right] + 7\,\mathrm{Im}\,\Omega \lrcorner (a_2 \wedge \mathrm{d}v_\perp) - 18 \left| \mathrm{d}A \lrcorner v \right|^2,$$

which gives exactly (5.4.8) when inserted into (5.3.4). This confirms that our formulas are correct. The bulk potential in this Calabi-Yau limit reads then

$$\begin{aligned}V_0 = \frac{1}{4\kappa^2}\int_M \mathrm{dvol}_7\, e^{4A}\Big\{&\left|\mathrm{d}\tilde{J}-*G\right|^2 - 3\left|\mathrm{d}\tilde{J}\lrcorner(v\wedge J)-6\,\mathrm{d}A\lrcorner v\right|^2 + \left|\mathrm{d}\tilde{v}\right|^2 + 40|a_2|^2 \\ &+ \left|\mathrm{Re}\,\mathrm{d}\tilde{\Omega}\lrcorner v\right|^2 + \left|\mathrm{d}\tilde{v}_\perp\lrcorner J\right|^2 - \left|8\,a_2 - \mathrm{d}\tilde{v}\lrcorner v\right|^2 + \left|4\,a_2 - \mathrm{d}\tilde{v}\lrcorner v + \mathrm{Im}\,\Omega\lrcorner\mathrm{d}\tilde{v}_\perp\right|^2 \\ &- \left|\tfrac{1}{4}\mathrm{Im}\big[\mathrm{d}\tilde{\Omega}\lrcorner(v\wedge\bar{\Omega})\big]\right|^2 - \left|\mathrm{Re}\,\mathrm{d}\tilde{\Omega}\lrcorner(v\wedge J)+4\,a_2 - \mathrm{d}\tilde{v}\lrcorner v + \mathrm{Im}\,\Omega\lrcorner\mathrm{d}\tilde{v}_\perp\right|^2\Big\}\,.\end{aligned} \quad (5.4.10)$$

As we have explained at the beginning of this section this potential has a BPS-like form. This becomes clear from (5.2.8) which states that for $W_1 = \ldots = W_5 = 0$ all torison classes except $\mathrm{Re}\,E$ and T_2 have to vanish under SUSY and that $\mathrm{Re}\,E = -3\mathrm{d}A\lrcorner v$. This implies that $a_2 = 0$ and that $\mathrm{d}\tilde{J}\lrcorner(v\wedge J) = 6\mathrm{d}A\lrcorner v$ under SUSY, respectively. So we see that all squares vanish for a supersymmetric setting. Furthermore, the only component of the four-form flux G which is not zero is $F^{2,2}$. This is in accordance with the necessity of a non-vanishing F in Horava-Witten theory. But it is also consistent with the fact that one expects zero H-flux once one reduces the theory to a heterotic compactification on a Calabi-Yau manifold. In order to see how this precisely works, we will consider the ten-dimensional limit in the next section.

5.4.3 The ten-dimensional limit

The most important consistency check of our previous results is the reduction of M-Theory to the heterotic string sector. As explained in section 2.3, the reduction is obtained by first performing the standard reduction of M-theory to type IIA theory as described e.g. in [9, 10], and then taking the limit $\pi\rho \to 0$ to move the two hyperplanes which are supporting the gauge fields on top of each other. This procedure should eventually lead to the results presented in the beginning of chapter 4.

We start by decomposing the eleven-dimensional metric as in (2.3.2) and then performing a compactification of the ten-dimensional space

$$\mathrm{d}s_{11}^2 = e^{-\frac{2}{3}\Phi}\mathrm{d}s_{10}^2 + e^{\frac{4}{3}\Phi}(\mathrm{d}x^{11}+X_1)^2 = e^{2A'-\frac{2}{3}\Phi}(\mathrm{d}s_4')^2 + e^{-\frac{2}{3}\Phi}(\mathrm{d}s_6')^2 + e^{\frac{4}{3}\Phi}(\mathrm{d}x^{11}+X_1)^2\,. \quad (5.4.11)$$

Here, Φ is the ten-dimensional dilaton and A' is the warp factor belonging to a compactification of string theory to four dimensions. g_4' denotes the metric of the emerging four-dimensional space, g_6' of the compact six-dimensional one, respectively, and X_1 is a one-form potential. Since we want to compare M-Theory compactifications to warped Minkowski space, we take g_4' to be the Minkowski metric. Comparing (5.4.11) with the previously defined metrics (2.4.1) and (2.4.5) we see that

$$2A' = 2A + \frac{2}{3}\Phi = 2A + \sigma, \qquad ds_7^2 = e^{-\frac{2}{3}\Phi}(ds_6')^2 + e^{\frac{4}{3}\Phi}(dx^{11} + X_1)^2, \qquad (5.4.12)$$

where we remind the reader that σ was the field used to describe the metric at the boundary in section 2.4.1. This means that the seven-dimensional space M splits into a six-dimensional base B and a one-dimensional fiber. Since locally every seven-dimensional SU(3) manifold can be decomposed into a six and a one-dimensional part, and since this one-dimensional piece is distinguished by v, we can also write

$$ds_7^2 = ds_6^2 + v \otimes v. \qquad (5.4.13)$$

Thus the metric g_6 that we used to construct the SU(3) structure and the metric g_6' appearing in (5.4.11) are related by $g_6 = e^{-2\Phi/3}g_6'$ which gives $J = e^{-2\Phi/3}J'$ and $\Omega = e^{-\Phi}\Omega'$. For the one-form v we get

$$v = e^{\frac{2}{3}\Phi}(dx_{11} + X_1). \qquad (5.4.14)$$

From the SUSY conditions for v (5.2.4a) it follows that

$$dA + \frac{1}{3}d\Phi = dA' = 0, \qquad\qquad dX_1 = 0 \qquad (5.4.15)$$

as it should be for the heterotic string. We also see that $dv \lrcorner v = W_0 = -2/3\, d\Phi_\perp = -d\sigma_\perp$, which justifies the choice we made for σ in section 5.3.2.

Since $(dA \lrcorner v)$ is not zero, these equations also imply that the dilaton does depend on the v-direction. This is not the case in the heterotic theory, and in order to see how this dependence vanishes, it is necessary to analyze the behavior of the flux F when the two hyperplanes are moved together. From equation (2.3.4) we know that the component of G_{11} having no leg along the reduced dimension becomes zero when one performs the dimensional reduction. But since we identify v with this direction, this means that $F = 0$ in this limit. For this reason, we conclude that we can also set $(dA \lrcorner v) \sim A_1$ to zero, once we go to the heterotic limit. A similar reasoning shows that $d\tilde{J} \lrcorner v = d\tilde{\Omega} \lrcorner v = 0$ for $\pi\rho \to 0$.

The supersymmetry conditions of section 4.1 can then be re-derived from our results (5.2.4)

$$\begin{aligned}
e^{-4A}d(e^{4A}J) &= *G = -*(v \wedge H) & &\Rightarrow & e^{-4A'+2\Phi}d(e^{4A'-2\Phi}J') &= *_6'H, & (5.4.16)\\
e^{-2A}d(e^{2A}J \wedge J) &= -2v \wedge G = 0 & &\Rightarrow & e^{-2A'+2\Phi}d(e^{2A'-2\Phi}J' \wedge J') &= 0,\\
e^{-3A}d(e^{3A}\Omega) &= 0 & &\Rightarrow & e^{-3A'+2\Phi}d(e^{3A'-2\Phi}\Omega') &= 0.
\end{aligned}$$

Note that in the first line one also has to rewrite the seven-dimensional Hodge star, $*(v \wedge H) = -e^{-2/3\Phi} *'_6 H$. This shows that our SUSY conditions are indeed compatible with the SUSY conditions found for string theory compactifications on six-dimensional SU(3) structure manifolds in section 4.1.

Due to the restrictions $\mathrm{d}A \lrcorner v = \mathrm{d}\tilde{J} \lrcorner v = \mathrm{d}\tilde{\Omega} \lrcorner v = 0$ in the heterotic limit, the linear piece (5.3.13) is identically zero and the bulk potential is simplified to

$$V_0 = \frac{1}{4\kappa_{10}^2} \int_B \mathrm{dvol}'_6 \, e^{4A'-2\Phi} \left\{ |\mathrm{d}\tilde{J}' - *'_6 H|^2 + |\mathrm{d}\tilde{\Omega}'|^2 - \frac{1}{4}|\mathrm{d}\tilde{\Omega}' \lrcorner J'^2|^2 + \frac{1}{4}|\mathrm{d}\tilde{J}'^2|^2 \right. \tag{5.4.17}$$
$$\left. - \left| \frac{1}{4} \mathrm{d}\tilde{J}'^2 \lrcorner J'^2 - \frac{1}{2} \mathrm{Re}(\mathrm{d}\tilde{\Omega}' \lrcorner \overline{\Omega}') \right|^2 + 4 |\mathrm{d}A'|^2 + 2|\mathrm{d}A'|^2 \right\} .$$

All squares appearing in this formula are taken with respect to the metric g'_6 in order to get the right factors of e^{Φ}, and we have absorbed the length of the eleventh dimension into the ten-dimensional coupling $\kappa^2 = 2\pi\rho \kappa_{10}^2$. This is indeed the action of heterotic supergravity to lowest order in α' that we presented in section 4.1.

The result for the boundary potential is even more easily obtained. Setting $3\sigma = 2\Phi$ and adding the contributions of the two boundaries gives

$$V_b = \frac{\alpha'}{8\kappa_{10}^2} \int_B \mathrm{dvol}'_6 \, e^{4A'-2\Phi} \left(\mathrm{Tr}|\mathcal{F}^{(6)}_{E_8 \times E_8} \lrcorner J|^2 + 2\,\mathrm{Tr}|(\mathcal{F}^{(6)}_{E_8 \times E_8})^{2,0}|^2 \right) \tag{5.4.18}$$
$$- \frac{\alpha'}{8\kappa_{10}^2} \int_B \mathrm{dvol}'_6 \, e^{4A'-2\Phi} \left(\mathrm{Tr}|R_+^{(6)'} \lrcorner J|^2 + 2\,\mathrm{Tr}|(R_+^{(6)'})^{2,0}|^2 \right)$$
$$- \frac{\alpha'}{8\kappa_{10}^2} \int_B \mathrm{dvol}'_6 \, e^{4A'-2\Phi} \left\{ 8 e^{-2A'} (\nabla_i \nabla_j e^{A'})(\nabla^i \nabla^j e^{A'}) - 4 |\mathrm{d}A' \lrcorner H|^2 - 12|\mathrm{d}A'|^4 \right\} .$$

This is the $\mathcal{O}(\alpha')$ result of section 4.1. We thus conclude that also in this limit our formulas provide the right results, since we obtain exactly the scalar potential of the heterotic string compactified on an SU(3) manifold.

We see therefore that the three limits that we considered in this section give evidence that our results for the scalar potential (5.3.19) and (5.3.22) are indeed correct. What does this then mean for the lift of DWSB vacua to M-theory?

5.5 DWSB in heterotic M-theory?

In this chapter we have shown so far that for a general compactification manifold with SU(3) structure it is not possible to obtain a BPS-like form of the potential. As we argued such a form would be mandatory in order to define a DWSB pattern also for heterotic M-theory. However, we have also seen, that under certain circumstances a BPS-like potential is achievable. As an example, we considered in the last section the case of a Calabi-Yau manifold fibered over a line. What does this then mean for the kind of DWSB that we discussed in chapter 4?

It should be clear from our analysis in section 5.3 that it is necessary for the BPS-like form that at least the two torsion classes $\operatorname{Re} S$ and T_2 are identically zero. On the other hand, the existence of the $(2,2)$-part of the flux component F is equally necessary in Horava-Witten theory. By (5.2.8) this means that the torsion class $\operatorname{Re} E$ is not allowed to vanish.

If we furthermore do not want to restrict the torsion classes $W_1, \ldots, W_5$ that appear also in ten-dimensional DWSB we are led to the following simple ansatz

$$\mathrm{d}J = \mathrm{d}J_\perp + \frac{2}{3}\operatorname{Re} E\, v \wedge J, \tag{5.5.1a}$$

$$\mathrm{d}\Omega = \mathrm{d}\Omega_\perp + \operatorname{Re} E\, v \wedge \Omega. \tag{5.5.1b}$$

By this we allow only compactification manifolds for which the torsion classes $\operatorname{Im} E$, V_2, T_2, and S are identically zero. Therefore, terms in the potential containing one of these classes will also vanish identically, and not only when SUSY is imposed. Note that for non-broken SUSY this means that F has to be $(2,2)$ and non-primitive, while for H all components are non-vanishing, as we had intended it.

Next we have to uplift our DWSB ansatz (4.2.1) and (4.2.3). Using the relations (5.4.16) obtained for the heterotic limit in the last section we consider the following uplift

$$\begin{aligned}
e^{2\Phi}\mathrm{d}\bigl(e^{-2\Phi}\,J'\wedge J'\bigr) &= 0 & &\longrightarrow & e^{-2A}\mathrm{d}\bigl(e^{2A}J\wedge J\bigr)_\perp &= 0, & &(5.5.2\mathrm{a})\\
e^{2\Phi}\mathrm{d}\bigl(e^{-2\Phi}\,J'\bigr) &= *'_6 H & &\longrightarrow & e^{-4A}\mathrm{d}\bigl(e^{4A}J\bigr)_\perp &= -*(v\wedge H), & &(5.5.2\mathrm{b})\\
\overline{\Omega}'\lrcorner\mathrm{d}\bigl(e^{-2\Phi}\Omega'\bigr) &= 0 & &\longrightarrow & \overline{\Omega}\lrcorner\mathrm{d}\bigl(e^{3A}\Omega\bigr) &= 0, & &(5.5.2\mathrm{c})\\
\mathrm{d}A' &= 0 & &\longrightarrow & \mathrm{d}\bigl(e^{2A}v\bigr) &= 0. & &(5.5.2\mathrm{d})
\end{aligned}$$

Here we adopted the same notation as in the last section, i.e. primed quantities are defined with respect to the heterotic theory. One should also note that (5.5.2b) implies the standard SUSY condition $e^{-4A}\mathrm{d}(e^{4A}J) = *G$ and that from (5.5.1b) and (5.5.2c) $\mathrm{d}(e^{3A}\Omega)\lrcorner v = 0$ follows.

Inserting all these conditions in the bulk potential (5.3.19) we find nearly the same result as for the heterotic case (4.2.4) up to a term proportional to $\operatorname{Im} W_1$

$$V'_0 = \frac{1}{4\kappa^2}\int_M \mathrm{dvol}_7\, e^{4A}\left\{\left|\mathrm{d}\tilde{\Omega}_\perp\right|^2 - \left|\mathrm{d}\tilde{\Omega}_\perp \wedge J\right|^2 - 4\operatorname{Im} W_1\bigl(12\,\mathrm{d}A\lrcorner v + 7\operatorname{Re} E\bigr)\right\}. \tag{5.5.3}$$

Again, this means that in the general case one can not simply lift the results of chapter 4 to M-theory. However, one should remember that we did also not try to solve the general case in chapter 4, but restricted us to the class of $\frac{1}{2}$DWSB that was characterized by the additional condition (4.4.1)

$$\text{Im}\left[e^{i\vartheta}\text{d}\left(e^{-2\Phi}\Omega\right)\right] = 0 ,$$

with an arbitrary phase ϑ. This condition implies that $\text{Im}(e^{i\vartheta}W_1) = 0$ and by further restricting the $\frac{1}{2}$DWSB case to $\vartheta = 0$ one gets $\text{Im}\,W_1 = 0$. For this subclass we find the result

$$V_0' = \frac{1}{4\kappa^2}\int_M \text{dvol}_7\, e^{4A}\left(\left|\text{d}\tilde{\Omega}_\perp\right|^2 - \left|\text{d}\tilde{\Omega}_\perp \wedge J\right|^2\right) , \tag{5.5.4}$$

that has exactly the same form as (4.2.4). Therefore, a simple lift of the $\frac{1}{2}$DWSB examples of section 4.5 is possible if one chooses $\vartheta = 0$ in (4.4.1). The resulting seven-dimensional manifold would then be an elliptically fibered four-dimensional Kähler space, that is fibered in addition over a line such that (5.5.1) holds. However, developing the details of such models, like e.g. the SUSY-breaking or compactification scales as was done for the heterotic string in chapter 4, would go far beyond the scope of this thesis.

Chapter 6

Conclusion

The supergravity ansatz to string theory is one of the most often used approaches to find physically correct compactifications to four dimensions. Its strength is that it allows to check the consistency of a given set of background fields explicitly. Therefore, it is possible to analyze a broad scope of phenomena with this approach, ranging from dualities to moduli stabilization. In our works [1–4], which we presented in this thesis, we were mostly concerned with the problem of supersymmetry breaking in heterotic $E_8 \times E_8$ supergravity and its lift to heterotic M-theory. We will end our presentation in this chapter by summarizing once more our main results and by giving an outlook for possible research, which could profit from our findings.

Summary of our results

We began this thesis by discussing the necessary supergravity theories in chapter 2. Although most of the contents of this chapter are standard, we would like to emphasize here again the non-trivial fact that it is possible to bring the action into the form of a four-dimensional integral over a potential, whose variation gives the same equations of motion as the variation of the full action. Only due to this fact it is feasible to consider only the potential instead of the action, which simplified our discussion.

In chapter 4 it was therefore possible to analyze supersymmetry breaking in heterotic $E_8 \times E_8$ supergravity starting from the effective potential. Furthermore, using G-structures we could rewrite the potential as well as the SUSY conditions in terms of the invariant forms J and Ω of the SU(3) structure of the compactification manifold. This showed that the potential can be brought into a BPS-like form, being a sum of squares of the SUSY conditions. Therefore, for a supersymmetric vacuum the potential vanishes and is automatically extremized.

Having obtained the conditions for unbroken SUSY, the next step was to find a controllable pattern of SUSY breaking. There, we used the technique of domain wall supersymmetry breaking

(DWSB) that was developed in the context of type II supergravity theories. For generic DWSB we found that NS5-branes which appear as domain walls in four dimensions are not any longer BPS-objects, while those branes which fill all four space-time dimensions, or which would appear as strings, still are BPS. This implied that it is still possible to define gauge bundles on the internal manifold, although it is not complex in the non-supersymmetric case. Furthermore, the potential is not automatically extremized once SUSY is broken. This result led to an additional EoM, which constrains the internal space, and restricts the SUSY breaking scale to lie below the compactification scale, respectively, as is necessary in order to have spontaneous SUSY breaking in four dimensions.

Solutions to the residual equation of motion could be found by studying a subclass of our vacua in section 4.4. Demanding an additional constraint we showed that for this class only compactifications on fibered manifolds with a two-dimensional fiber and a four-dimensional base are allowed. From a four-dimensional perspective these vacua can be identified as no-scale models. In these models the potential does not depend on a given set of fields. Therefore, these fields can develop SUSY breaking F-terms which will not alter the minimum of the potential and thus do not endanger the stability of the solution. On the other hand, moduli coming from these fields are not stabilized. For our ansatz we showed that it are the pseudo-Kähler moduli of the base space which develop F-terms and which are thus responsible for SUSY breaking.

Considering in this context elliptic fibrations of K3 as the simplest examples, we could demonstrate that the dilaton gets stabilized by the Bianchi identity of H, and that the size of the elliptic fiber gets fixed due to flux quantization.

Since gaugino condensation is the typical way to break SUSY in heterotic string theory, we included a fermionic condensate in our discussion at the end of chapter 4. We were particularly interested in the question whether one could restore SUSY by having both, non-vanishing flux and a non-vanishing condensate, respectively. As it turned out this is not possible, and it seems as if the two ways to break supersymmetry are orthogonal to each other.

After the quite elaborate study of the weakly coupled heterotic string, we wanted to discuss in chapter 5 the possibilities of lifting our solutions to the strongly coupled theory, i.e. to elevendimensional supergravity with two ten-dimensional boundaries. Also for this theory we could show in chapter 2 the equivalence of the EoM's derived from the full action and from an effective potential. Therefore, this potential provided again the right starting point for the analysis.

However, it was very difficult to answer the question whether the potential could be given in a BPS-like form as in the ten-dimensional case. After tedious calculations that we presented in chapter 5, we came to the conclusion that this is not possible for a generic seven-dimensional internal manifold, but only when certain torsion classes are identically set to zero. Since this is quite surprising, we performed several crosschecks of our formulas going to the limits of G_2 holonomy,

SU(3) holonomy, and the ten-dimensional limit. In all these checks our equations provided the correct results.

Having no BPS-like potential seems at first sight like the end of DWSB in the strongly coupled heterotic string. But considering only the explicit examples, which we constructed in chapter 4, we found that these examples can be lifted to heterotic M-theory, and thus could provide controllable non-SUSY vacua for the strongly coupled heterotic string. Although we could not analyze this in more detail in our works, it could definitively be a viable direction of further research.

Further research

Although we think that we could give a compact picture of DWSB in the last chapters there are of course several loose ends, that one could treat in future projects.

One of them that we already alluded to in section 4.3 is the mathematical issue of the pseudo-HYM condition on the field strength. In the complex case it can be shown that this condition is equivalent to a bundle stability criterion. One important ingredient of this theorem is that the space is at least balanced, which corresponds in our setting to the possible presence of BPS-strings. Since we do preserve this condition also in the non-supersymmetric case, it might be interesting to study this subject again from the perspective of calibrated submanifolds, where one might hope to find generalizations of the original Donaldson-Uhlenbeck-Yau theorem.

A more physical line of research should be concerned with the stabilization of the remaining moduli of our examples. As we explained the no-scale structure of the model intrinsically forbids that all moduli can be stabilized with fluxes. In this case, one normally invokes non-perturbative effects to give masses to these moduli. But since we already included such effects studying gaugino condensation, one should redo our four-dimensional analysis in the presence of a condensate, and check whether one can stabilize the pseudo-Kähler moduli of the base, too.

Otherwise, one could also leave our simple example class and search for totally new compactification manifolds that satisfy the general DWSB conditions but not the restricted $\frac{1}{2}$DWSB constraints. Eventually, this could be achieved by considering toric varieties [246] along the lines of [247]. The main difficulty here would be to translate our requirements on torsion classes into the language of toric varieties that was used in [247]. But maybe one could get by this method large enough samples of compactification manifolds to test whether it is possible to obtain realistic four-dimensional physics, i.e. to stabilize the moduli, to include a consistent gauge bundle, to satisfy the Bianchi identity and the flux quantization, and to have the right SUSY breaking scale. From the viewpoint of four-dimensional physics this would be essential, although we know that it is extremely hard to satisfy all these conditions in one single model.

One should also notice that recently there have been attempts [203] to place the heterotic string in the broader context of doubled field theory [16, 17] and generalized geometry [248, 249]. It is of course interesting whether and how DWSB fits into this framework, which provides a unified description of both T-dual heterotic theories.

Besides these topics, which are connected to the weakly coupled heterotic string, one can of course also pursue the strongly coupled description. The question, which comes into ones mind immediately, is of course how it is possible that one cannot obtain a BPS-like form of the potential for a generic seven-dimensional SU(3) structure manifold. As we have seen in section 5.4.3 this is not a problem for the weakly coupled string, since the correct form of the potential and the SUSY conditions in ten dimension can be deduced from any starting point in eleven dimensions. It seems therefore that the BPS-like structure vanishes due to strong coupling effects, which definitively should be analyzed further.

Concentrating on the phenomenological aspects of heterotic M-theory one should ask which implications the lift of our $\frac{1}{2}$DWSB examples to M-theory would have. First, one should try to recapitulate all steps of section 4.4 and section 4.5 for the strongly coupled theory and compare the results to the weakly coupled case. Moreover, it is expected that on an intermediate energy scale the eleventh dimension is still non-compact while six dimensions are already compactified. As this could be just in the range of the SUSY breaking scale, and as strong coupling effects are still visible then, one should also study the resulting effective five-dimensional theory. Then, one could also include a gaugino condensate on the two four-dimensional boundaries of the five-dimensional theory and analyze the interplay of the SUSY breaking coming from DWSB and the condensate. Whether this could lead to four-dimensional physics that resembles the standard model is of course the ultimate question.

Appendix A

Appendix

A.1 Conventions

Indices

There are various sorts of indices appearing in the thesis. An underline will be used to distinguish flat from curved indices when needed. We denote with $M, N, \ldots$ eleven dimensions and with $I, J, \ldots$ ten dimensions (either the ten of heterotic supergravity, or the first ten in M-theory). $m, n, \ldots$ will be used for seven internal dimensions and $i, j, \ldots$ for six, respectively (either the six dimensions of an internal manifold of a heterotic compactification, or the first six of an M-theory compactification). Letters from the beginning of the alphabet $A, B, \ldots$ $(a, b, \ldots)$ will be used in formulas that hold in eleven (seven) as well as in ten (six) dimensions. The Greek letters $\mu, \nu \ldots$ will stand for the four-dimensional external space-time.

Gamma matrices of SO(1, 10)

Our conventions on gamma matrices in eleven dimensions are as follows. With Γ^N we denote gamma matrices of $SO(1, 10)$ which are 32×32 matrices. We split the Γ^N according to

$$\Gamma^\mu = \gamma^\mu \otimes \mathbb{1} = e^{-A}\hat{\gamma}^\mu \otimes \mathbb{1}, \qquad \Gamma^m = \gamma_{(4)} \otimes \gamma^m. \tag{A.1.1}$$

The γ^μ are taken to be real and symmetric 4×4 matrices and represent the gamma matrices of the warped four-dimensional external spacetime. The γ^m are purely imaginary, antisymmetric, and 8×8, which implies that the Γ^N are real and symmetric. The four-dimensional chirality operator is defined as

$$\gamma_{(4)} = i\gamma^{\underline{0}}\gamma^{\underline{1}}\gamma^{\underline{2}}\gamma^{\underline{3}}. \tag{A.1.2}$$

From $\Gamma^{\underline{10}}\Gamma^{\underline{10}} = \mathbb{1}$ and $\Gamma^{\underline{10}} = \Gamma^{\underline{0}} \cdot \ldots \cdot \Gamma^{\underline{9}}$ it follows that

$$\gamma^{\underline{10}}\gamma^{\underline{10}} = -i\,\gamma^{\underline{4}}\ldots\gamma^{\underline{10}} = \mathbb{1} \ . \tag{A.1.3}$$

An explicit representation of these gamma matrices can be found as in [23] and is given by

$$\gamma^{\underline{0}} = \begin{pmatrix} 0 & \sigma^1 \\ -\sigma^1 & 0 \end{pmatrix}, \gamma^{\underline{1}} = \begin{pmatrix} 0 & \sigma^1 \\ \sigma^1 & 0 \end{pmatrix}, \gamma^{\underline{2}} = \begin{pmatrix} \sigma^3 & 0 \\ 0 & \sigma^3 \end{pmatrix}, \gamma^{\underline{3}} = \begin{pmatrix} -\sigma^1 & 0 \\ 0 & \sigma^1 \end{pmatrix}, \tag{A.1.4}$$

$$\gamma^{\underline{4,5,6}} = \begin{pmatrix} \alpha^{1,2,3} & 0 \\ 0 & \alpha^{1,2,3} \end{pmatrix}, \gamma^{\underline{7,8,9}} = \begin{pmatrix} 0 & \beta^{1,2,3} \\ \beta^{1,2,3} & 0 \end{pmatrix}, \gamma^{\underline{10}} = \begin{pmatrix} 0 & -i\mathbb{1}_4 \\ i\mathbb{1}_4 & 0 \end{pmatrix}.$$

With σ^i the Pauli-matrices, the 4×4 matrices α^i, β^j are given by

$$\alpha^1 = \begin{pmatrix} 0 & i\sigma^1 \\ -i\sigma^1 & 0 \end{pmatrix}, \quad \alpha^2 = \begin{pmatrix} 0 & -i\sigma^3 \\ i\sigma^3 & 0 \end{pmatrix}, \quad \alpha^3 = \begin{pmatrix} -\sigma^2 & 0 \\ 0 & -\sigma^2 \end{pmatrix},$$

$$\beta^1 = \begin{pmatrix} 0 & -\sigma^2 \\ -\sigma^2 & 0 \end{pmatrix}, \quad \beta^2 = \begin{pmatrix} 0 & i\mathbb{1} \\ -i\mathbb{1} & 0 \end{pmatrix}, \quad \beta^3 = \begin{pmatrix} \sigma^2 & 0 \\ 0 & -\sigma^2 \end{pmatrix}. \tag{A.1.5}$$

Defining $\gamma^{m_1\ldots m_n} = \gamma^{[m_1}\ldots\gamma^{m_n]}$ the relation between the antisymmetrized product of n and $7-n$ gamma matrices is

$$\gamma^{m_1\ldots m_n} = i\,(-1)^{1+n(n-1)/2} \frac{1}{(7-n)!}\,\epsilon^{m_1\ldots m_n}{}_{m_{n+1}\ldots m_7}\gamma^{m_{n+1}\ldots m_7} \ . \tag{A.1.6}$$

Due to our manifestly real gamma matrices, the Majorana condition on an 11-dimensional spinor ϵ simply reads

$$\epsilon^* = \epsilon \ . \tag{A.1.7}$$

Gamma matrices of SO(1,9)

In ten dimensions we choose again a real representation for the $SO(1,9)$ gamma matrices Γ^I. Then we can split the 32×32 matrices Γ^I in terms of real symmetric four- and pure imaginary antisymmetric six-dimensional gamma matrices $\hat{\gamma}^\mu$ (associated with the unwarped X_4 metric) and γ^i, analogously to the $SO(1,10)$ case

$$\Gamma^\mu = e^{-A}\hat{\gamma}^\mu \otimes \mathbb{1} \ , \qquad \Gamma^i = \gamma_{(4)} \otimes \gamma^i \ . \tag{A.1.8}$$

Since we use a real representation of the ten-dimensional gamma matrices Γ^I, Majorana spinors of $SO(1,9)$ will again be real. The ten-dimensional chirality operator is then defined to be

$$\Gamma_{(10)} = \Gamma^{\underline{0...9}}. \quad (A.1.9)$$

The heterotic ten-dimensional supersymmetry generator ϵ is Majorana-Weyl, meaning that it is real and it satisfies the chirality condition $\Gamma_{(10)}\epsilon = \epsilon$. The four- and six-dimensional chiraltiy operators are defined by

$$\Gamma_{(10)} = \gamma_{(4)} \otimes \gamma_{(6)}, \quad (A.1.10)$$

and are given explicitly by

$$\gamma_{(4)} = i\gamma^{\underline{0}}\gamma^{\underline{1}}\gamma^{\underline{2}}\gamma^{\underline{3}}, \quad \text{and} \quad \gamma_{(6)} = -i\gamma^{\underline{4}}\gamma^{\underline{5}}\gamma^{\underline{6}}\gamma^{\underline{7}}\gamma^{\underline{8}}\gamma^{\underline{9}}. \quad (A.1.11)$$

The relation between the antisymmetrized product of n $SO(6)$ gamma matrices and $(6-n)$ matrices is given by

$$\gamma^{i_1...i_n} = i(-1)^{1+n(n-1)/2}\frac{1}{(6-n)!}\epsilon^{i_1...i_n}{}_{i_{n+1}...i_6}\gamma^{i_{n+1}...i_6}\gamma_{(6)}. \quad (A.1.12)$$

Contractions

A slash will denote normalized antisymmetrized contraction of a p-form A_p with gamma matrices

$$\slashed{A}_p = \frac{1}{p!}\gamma^{A_1...A_p}(A_p)_{A_1...A_p}, \quad (A.1.13)$$

while $\lrcorner$ is used to contract p- and q-forms

$$A_p \lrcorner B_q = \frac{1}{(p-q)!\,q!}A_{A_1...A_qB_1...B_{p-q}}B^{A_1...A_q}dx^{B_1...B_{p-q}}. \quad (A.1.14)$$

$|A_p|^2$ means total contraction of a p-form A_p with its complex conjugate

$$|A_p|^2 = \frac{1}{p!}A_{A_1...A_p}\overline{A}^{A_1...A_p}. \quad (A.1.15)$$

Hodge star

We will be concerned with Hodge stars in several dimensions. These will always be defined in the standard way

$$*_d A_p = \frac{\sqrt{|g_d|}}{p!(d-p)!} \epsilon_{A_1...A_{d-p}}{}^{A_{d-p+1}...A_d} (A_p)_{A_{d-p+1}...A_d} dx^{A_1...A_{d-p}} \qquad (A.1.16)$$

$$= \frac{1}{p!(d-p)!} \epsilon_{\underline{A_1}...\underline{A_{d-p}}}{}^{\underline{A_{d-p+1}}...\underline{A_d}} (A_p)_{\underline{A_{d-p+1}}...\underline{A_d}} dx^{\underline{A_1}...\underline{A_{d-p}}} .$$

$$(A.1.17)$$

Most often we will encounter the Hodge star in six and seven (euclidean) dimensions. There, one has the useful identities

$$*_6^2 A_p = (-1)^p A_p , \qquad *_7^2 A_p = A_p , \qquad (A.1.18)$$

and

$$A_p \wedge *_6 B_p = (-1)^p A_p \lrcorner B_p , \qquad A_p \wedge *_7 B_p = A_p \lrcorner B_p . \qquad (A.1.19)$$

Spinor decomposition

We will decompose the $SO(1,10)$ Majorana spinor ϵ_{11} of Horava-Witten theory as

$$\epsilon_{11} = \chi_+ \otimes \eta_+ + \chi_- \otimes \eta_- = \chi_+ \otimes \eta_+ + c.c. . \qquad (A.1.20)$$

Here, χ_+ is an anticommuting four-dimensional spinor of positive chirality, while η_+ is a commuting $SO(7)$-spinor, that reduces to a positive $SO(6)$ spinor at the boundary and after dimensional reduction to heterotic supergravity.

In the same sense the $SO(1,9)$ spinor ϵ_{10} of heterotic supergravity decomposes as

$$\epsilon_{10} = \zeta \otimes \eta_+ + c.c. . \qquad (A.1.21)$$

Again, ζ is an anticommuting four-dimensional spinor of positive chirality. The spinor η_+ is a commuting $SO(6)$-spinor, also of positive chirality. We give the same name to the two internal spinors of the eleven- and the ten-dimensional case for purpose, since they get identified in the course of dimensional reduction.

The four-dimensional spinors are supposed to satisfy a Killing spinor equation in AdS-space relating them to the AdS radius $R_{\text{AdS}} = \frac{1}{|w_0|}$. For χ_+ (and analogously for ζ) this reads

$$\hat{\nabla}_\mu \chi_+ = \frac{1}{2} w_0^* \hat{\gamma}_\mu \chi_- . \qquad (A.1.22)$$

A.2 Dual formulation of heterotic supergravity

Here, we give a short introduction to the dual formulation of heterotic supergravity as needed in chapter 4. The dual formulation of the heterotic theory is expressed in terms of the seven-form flux $\hat{H} = e^{-2\phi} * H$. In this formulation the six-form potential $\hat{B}$, $d\hat{B} = \hat{H}$, plays the role of the fundamental field and couples electrically to the NS5-brane charge of the background, and the BI (2.2.5) arises as the equation of motion of $\hat{B}$. For this reason, this frame is the natural one to describe the coupling of NS5-branes.

The complete dual formulation up to order α' can be found in [200], and the dualization procedure relating the two formulations is discussed in detail in [250]. Here, we just focus on the bosonic sector. One starts with the action

$$S' = S - \frac{1}{2\kappa_{10}^2} \int_{X_{10}} \hat{B} \wedge \left[dH + \frac{\alpha'}{4}(\operatorname{Tr} F \wedge F - \operatorname{Tr} R_+ \wedge R_+) \right], \tag{A.2.1}$$

where S has the same form as in (2.2.1), and where H and $\hat{B}$ should be considered as elementary independent fields. By varying with respect to $\hat{B}$ one gets the Bianchi identity (2.2.5) and then, integrating out $\hat{B}$ just produces the original action (2.2.1).

On the other hand, by extremizing S' with respect to H, one gets

$$\hat{H} = e^{-2\phi} * H. \tag{A.2.2}$$

By plugging (A.2.2) into (A.2.1) and keeping only $\mathcal{O}(\alpha')$ terms, we arrive at the dual action

$$\begin{aligned}\hat{S} &= \frac{1}{2\kappa_{10}^2} \int d^{10}x \sqrt{-g}\, e^{-2\phi} \left[\mathcal{R} + 4(d\phi)^2 - \frac{1}{2} e^{4\phi} \hat{H}^2 + \frac{\alpha'}{4}(\operatorname{Tr} R_+^2 - \operatorname{Tr} F^2) \right] \\ &\quad - \frac{\alpha'}{8\kappa^2} \int_{X_{10}} \hat{B} \wedge (\operatorname{Tr} F \wedge F - \operatorname{Tr} R_+ \wedge R_+). \end{aligned} \tag{A.2.3}$$

The supersymmetry transformations in the dual formulation are as in (2.2.6), up to terms which vanish on-shell at order α'.

Let us also consider the duality transformation in presence of a non-vanishing gaugino condensate. It is useful to introduce a three-form Σ defined by

$$\Sigma_{IJK} = \frac{\alpha'}{4} \operatorname{Tr} \bar{\chi} \Gamma_{IJK} \chi. \tag{A.2.4}$$

In the ordinary formulation which uses the 3-form H as fundamental, the relevant terms in the action are now as in (4.6.1). By performing the duality transformation described above we get

$$\hat{H} = e^{-2\phi} * T, \qquad \text{with} \qquad T = H - \frac{1}{2}\Sigma, \tag{A.2.5}$$

and the dual action with non-vanishing gaugino terms reads

$$\hat{S}' = \hat{S} - \frac{\alpha'}{4\kappa_{10}^2} \int d^{10}x \sqrt{-g}\, e^{-2\phi}\, \bar{\chi}(\slashed{D} - \frac{1}{4}\slashed{T})\chi\,. \tag{A.2.6}$$

Now the supersymmetry transformations are modified at $\mathcal{O}(\alpha')$ by the presence of $\Sigma \neq 0$ and take the form (4.6.7).

A.3 Supersymmetry breaking in the presence of a gaugino condensate

Here, we discuss how the supersymmetry breaking equations alter if one allows a gaugino condensate. The supersymmetry variations of the gravitino (2.2.6a) and dilatino (2.2.6b) get changed into

$$\delta\psi_I = \left(\nabla_I - \frac{1}{4}\slashed{H}_I + \frac{1}{16}\slashed{\Sigma}\Gamma_I\right)\epsilon, \tag{A.3.1a}$$

$$\delta\lambda = \left(\slashed{\partial}\phi - \frac{1}{2}\slashed{H} - \frac{1}{8}\slashed{\Sigma}\right)\epsilon, \tag{A.3.1b}$$

while the variation of the gaugino (2.2.6c) remains unchanged. For the external component of (A.3.1a) one obtains then

$$\delta\psi_\mu = \frac{1}{2}e^A \hat{\gamma}_\mu \zeta \otimes \left(\slashed{\partial}A\eta_+ + e^{-A}w_0 \eta_+^* - \frac{1}{8}\slashed{\Sigma}\eta_+\right) + \text{c.c.}\,. \tag{A.3.2}$$

This shows that after a gaugino condensate is added the condition $\delta\psi_\mu = 0$ no longer forces the cosmological constant to be zero or the warp factor to be constant. Allowing for additional violation of $\delta\psi_\mu = 0$ yields

$$\slashed{\Sigma}\eta_+ = 8\slashed{\partial}A\eta_+ - 16\,c_i\gamma^i\eta_+ + 8\,e^{-A}w_0\eta_+^* - 16\,h\eta_+^* = 8\slashed{\partial}A\eta_+ - 16\,c_i\gamma^i\eta_+ + 8\,e^{-A}\sigma_0\eta_+^*, \tag{A.3.3}$$

where h and c_i measure the SUSY-breaking and $\sigma_0 = w_0 - 2e^A h$. c_i is restricted by $P_i{}^j c_j = 0$. Below, we will impose that $c_i = 0$, so that $\delta\psi_\mu \propto \zeta \otimes \eta_+^* + \text{c.c.}$, which is a natural assumption if we want to interpret the SUSY-breaking in $\mathcal{N} = 1$ four-dimensional terms.

The internal component of the gravitino variation and the dilatino variation read

$$\delta\psi_i = \zeta \otimes \left(\nabla_i - \frac{1}{4}\slashed{H}_i + \frac{1}{16}\slashed{\Sigma}\Gamma_i\right)\eta_+ + c.c,\qquad\text{(A.3.4a)}$$

$$\delta\lambda = \zeta \otimes \left(\slashed{\partial}\phi - \frac{1}{2}\slashed{H} - \frac{1}{8}\slashed{\Sigma}\right)\eta_+ + c.c,\qquad\text{(A.3.4b)}$$

which can be decomposed in analogy with (4.2.12) as

$$\left(\nabla_i - \frac{1}{4}\slashed{H}_i + \frac{1}{16}\slashed{\Sigma}\Gamma_i\right)\eta_+ = i\,\tilde{p}_i\eta_+ + \tilde{q}_{ij}\gamma^j\eta_+^*,\qquad\text{(A.3.5a)}$$

$$\left(\slashed{\partial}\phi - \frac{1}{2}\slashed{H} - \frac{1}{8}\slashed{\Sigma}\right)\eta_+ = \tilde{u}_i\gamma^i\eta_+ + \tilde{r}\eta_+^*,\qquad\text{(A.3.5b)}$$

where $\tilde{p}_i$ is real and $\tilde{q}_{ij}$ and $\tilde{u}_i$ are restricted by the projector conditions $\bar{P}_j{}^k\tilde{q}_{ik} = P_i{}^k\tilde{u}_k = 0$. Hence, the violation of $\delta\psi_i = \delta\lambda = 0$ can be expressed by the parameters $\tilde{p}_i$, $\tilde{q}_{ij}$, $\tilde{u}_i$ and $\tilde{r}$. The exterior derivatives of the SU(3) structure tensors J and Ω read then

$$e^{-2A+2\phi}\mathrm{d}\left(e^{2A-2\phi}J\wedge J\right) = 4\,\mathrm{Re}\,(v-\tilde{u})\wedge J\wedge J - 8\,\mathrm{Re}\,(\tilde{s}^*\wedge\Omega),\qquad\text{(A.3.6a)}$$

$$e^{-3A+2\phi}\mathrm{d}\left(e^{3A-2\phi}\Omega\right) = (2\,i\,\tilde{p} - 2\,u + 5\,c)\wedge\Omega$$
$$- \left(\tilde{r} + e^{-A}w_0 - 2h\right)J\wedge J + 8\,i\,\tilde{s}\wedge J,\qquad\text{(A.3.6b)}$$

$$e^{-4A+2\phi}\mathrm{d}\left(e^{4A-2\phi}J\right) - *T = \mathrm{Im}\left(\left[2\,\tilde{r}^* + 3\,e^{-A}\bar{w}_0 - 6\,h^*\right]\Omega\right) +$$
$$\mathrm{Re}\,(14\,c - 4\,\tilde{u}')\wedge J - 2\,\mathrm{Im}\left(\tilde{t}^{*n}\wedge\iota_n\Omega\right).\qquad\text{(A.3.6c)}$$

Here we used $\tilde{s} = \frac{1}{2}\tilde{q}_{ij}\mathrm{d}y^i\wedge\mathrm{d}y^j$, $\tilde{t}^j = \tilde{q}_i{}^j\mathrm{d}y^i$, $\tilde{u} = \tilde{u}_i\mathrm{d}y^i$ and $\tilde{p} = \tilde{p}_i\mathrm{d}y^i$. After setting $\tilde{p}_i = \tilde{q}_{ij} = \tilde{u}_i = c_i = \tilde{r} = h = 0$ one obtains (4.6.17) and (4.6.18).

As in the gaugino-less case the parameters $\tilde{p}_i$, $\tilde{q}_{ij}$, $\tilde{u}_i$, c_i, $\tilde{r}$, and h get severely restricted by our SUSY-breaking ansatz. First, we impose $c = 0$, for the reason discussed above. Then, by imposing (4.6.17a) and (4.6.18) one obtains $\tilde{u} = 0$, $\tilde{s}^{2,0} = 0$, $\tilde{r} - 3h = g^{ij}q_{ij}$, and $\tilde{q}^{2,0} = 0$. Furthermore, from (4.2.3) one gets $\tilde{p} = 0$. Hence, the remaining SUSY-breaking condition is

$$e^{-3A+2\phi}\mathrm{d}\left(e^{3A-2\phi}\Omega\right) = -\left(\tilde{r} + e^{-A}\sigma_0\right)J\wedge J + 8\,i\,\tilde{s}\wedge J = -\hat{r}J\wedge J + 8\,i\,\tilde{s}\wedge J,\qquad\text{(A.3.7)}$$

which is of the same form as (4.2.19). Thus, $\tilde{s} = -\frac{i}{6}\hat{r}J + s_\mathrm{P}$ and the comparison to (4.6.24) gives

$$\hat{r} = 3W_1, \qquad, \quad s_\mathrm{P} = -\frac{i}{8}W_2.\qquad\text{(A.3.8)}$$

A.4 The scalar curvature of G_2 structure manifolds

In this section we would like to show how to obtain (5.1.3) from the results of [210]. This appendix is in no way self contained, and the reader should consult [210] for more details. We start with equation (4.27) of [210], which reads $R = 6\,\phi_{mnp}\,T^{mnp}$, and can be rewritten as[1]

$$R = 6\,\phi_{mnp}\,T^{mnp} = A\,\delta\tau_1 + B\,\tau_0^2 + C\,|\tau_1|^2 + D\,|\tau_2|^2 + E\,|\tau_3|^2 \ . \tag{A.4.1}$$

The only two quantities not given explicitly in [210] are T_{mnp} and $\delta\tau_1$. To obtain T_{mnp} one has to calculate the covariant derivative of the torison τ using the formulas (4.9) and (4.19) of [210]

$$\begin{aligned}
D\tau_n &= \mathrm{d}\tau_n + \theta_n{}^m \wedge \tau_m - \phi_n{}^{mp}\tau_p \wedge \tau_m \tag{A.4.2}\\
&= (\mathrm{d}T_n{}^m) \wedge \omega_m + T_n{}^m\,\mathrm{d}\omega_m + \theta_n{}^m \wedge \tau_m - \phi_n{}^{mp}\tau_p \wedge \tau_m \\
&= (\mathrm{d}T_n{}^m - T_n{}^p\theta_p{}^m + T_p{}^m\theta_n{}^p) \wedge \omega_m - (2\phi_r{}^{mq}T_n{}^r T_q{}^p + \phi_n{}^{rq}T_q{}^p T_r{}^m)\omega_p \wedge \omega_m \\
&= (S_n{}^{mp} - 2\phi_r{}^{mq}T_n{}^r T_q{}^p - \phi_n{}^{rq}T_q{}^p T_r{}^m)\omega_p \wedge \omega_m \\
&= \frac{1}{2}T_n{}^{mp}\,\omega_p \wedge \omega_m
\end{aligned}$$

and hence

$$T_{mnp} = -2\,S_{mnp} - 4\,\phi_{qpr}\,T_m{}^q T^r{}_n - 2\,\phi_{mqr}\,T_p^q\,T^r{}_n \ . \tag{A.4.3}$$

From this we obtain for R

$$\begin{aligned}
R &= -12\,\phi^{mnp}(S_{mnp} + 2\,\phi_{qpr}\,T_m{}^q T^r{}_n + \phi_{mqr}\,T_p^q\,T^r{}_n) \tag{A.4.4}\\
&= -12\,\phi^{mnp}\,S_{mnp} + 36\,(T_n{}^n)^2 + 12\,\psi^{mnpq}\,T_{mn}T_{pq} - 24\,T^{mn}T_{mn} - 12\,T^{mn}T_{nm} \ .
\end{aligned}$$

For $\delta\tau_1$ we get

$$\delta\tau_1 = -*\mathrm{d}*\tau_1 = -\phi^{mnp}S_{mnp} - 2\,\psi^{mnpq}T_{mn}T_{pq} - 2\,T^{mn}(T_{mn} - T_{nm}) \ . \tag{A.4.5}$$

The torsion classes, defined through

$$\mathrm{d}\phi = \tau_0\,\psi + 3\,\tau_1 \wedge \phi + *\tau_3 \ , \quad \mathrm{d}\psi = 4\,\tau_1 \wedge \psi + \tau_2 \wedge \phi \ , \tag{A.4.6}$$

---

[1] We changed the ϵ-notation of [210] such that $\epsilon_{mnp} = \phi_{mnp}$ and $\epsilon_{mnpq} = \psi_{mnpq}$.

are given by

$$\tau_0 = \frac{1}{7} d\phi \lrcorner \psi = \frac{24}{7} T_n{}^n,$$ (A.4.7)

$$\tau_1 = -\frac{1}{12} d\phi \lrcorner \phi = \frac{1}{12} d\psi \lrcorner \psi = \phi^{mnp} T_{mn} \omega_p,$$

$$\tau_2 = \frac{1}{2}(d\psi \lrcorner \phi - *d\psi) - 2\tau_1 \lrcorner \phi = -*d\psi + 4\tau_1 \lrcorner \phi = (-\psi^{mnpq} T_{mn} + 4T^{pq}) \omega_p \wedge \omega_q,$$

$$\tau_3 = *d\phi - \tau_0 \phi + 3\tau_1 \lrcorner \psi = \left(\frac{3}{7} T_q{}^q \phi^{mnp} - \frac{3}{2} \phi^{qnp}(T_q{}^m + T^m{}_q)\right) \omega_m \wedge \omega_n \wedge \omega_p.$$

The squares of these are

$$\tau_0^2 = \frac{576}{49}(T_n{}^n)^2,$$ (A.4.8)

$$|\tau_1|^2 = \psi^{mnpq} T_{mn} T_{pq} + T^{mn}(T_{mn} - T_{nm}),$$

$$|\tau_2|^2 = -12\,\psi^{mnpq} T_{mn} T_{pq} + 24\,T^{mn}(T_{mn} - T_{nm}),$$

$$|\tau_3|^2 = -\frac{72}{7}(T_n{}^n)^2 + 36\,T^{mn}(T_{mn} + T_{nm}).$$

Comparing terms containing S_{mnp}, $(T_n{}^n)^2$, $\psi^{mnpq} T_{mn} T_{pq}$, $T^{mn} T_{mn}$, and $T^{mn} T_{nm}$, respectively, in (A.4.1) and (A.4.4) one gets equation (4.28) of [210]

$$R = 12\,\delta\tau_1 + \frac{21}{8}\tau_0^2 + 30\,|\tau_1|^2 - \frac{1}{2}|\tau_2|^2 - \frac{1}{2}|\tau_3|^2.$$ (A.4.9)

Using (4.16) of [210], $d\phi = \psi^{mnpq} T_m \wedge \omega_n \wedge \omega_p \wedge \omega_q$ and $d\psi = -6\tau^p \wedge \omega_p \wedge \phi$, it is possible to show that

$$|d\phi|^2 = 36\left[2(T_n{}^n)^2 + 2T^{mn} T_{mn} + \psi^{mnpq} T_{mn} T_{pq}\right],$$ (A.4.10)

$$|d\psi|^2 = 36\left[2T^{mn}(T_{mn} - T_{nm}) + \psi^{mnpq} T_{mn} T_{pq}\right].$$

From these expressions we find

$$|\tau_2|^2 = |d\psi|^2 - 48\,|\tau_1|^2,$$ (A.4.11)

$$|\tau_3|^2 = |d\phi|^2 - 36\,|\tau_1|^2 - 7\,|\tau_0|^2,$$

leading to the final expression for the scalar curvature

$$\begin{aligned}R &= 12\,\delta\tau_1 + \frac{49}{8}\tau_0^2 + 72\,|\tau_1|^2 - \frac{1}{2}|\mathrm{d}\phi|^2 - \frac{1}{2}|\mathrm{d}\psi|^2 \\ &= -\nabla^m\left(\mathrm{d}\psi\lrcorner\psi\right)_m + \frac{1}{2}|\mathrm{d}\psi\lrcorner\psi|^2 + \frac{1}{8}|\mathrm{d}\phi\lrcorner\psi|^2 - \frac{1}{2}|\mathrm{d}\phi|^2 - \frac{1}{2}|\mathrm{d}\psi|^2 \ .\end{aligned} \quad (A.4.12)$$

A.5 SUSY constraints

In this section we give the full list of constraints coming from the external SUSY variation (5.2.3a). In these tables 'Ext' stands for equation (5.2.3a) and $\gamma_{[n]}$ denotes n antisymmetrized gamma matrices. We found the mathematica package GAMMA [251] very useful for the calculation.

Table A.1: Constraints from $\delta\Psi_\mu = 0$ coming from $\eta_+^T \gamma_{[n]} \text{Ext} + \text{Ext}^T \gamma_{[n]} \eta_+$.

$144 e^{-A} w_0 \Sigma_0 + (\tilde{\Sigma}_4)_{l_1 l_2 l_3 l_4} G^{l_1 l_2 l_3 l_4} = 0$

$4 i \mu (\tilde{\Sigma}_3)_{mnp} + 12 e^{-A} w_0 (\Sigma_3)_{mnp} = 12 (\tilde{\Sigma}_4)_{l_1 mnp} \partial^{l_1} A + 3 (\tilde{\Sigma}_3)_{a_1 a_2 [m} G^{a_1 a_2}{}_{np]}$

$2 i \mu (\tilde{\Sigma}_4)_{mnpq} + 6 e^{-A} w_0 (\Sigma_4)_{mnpq} + 24 (\tilde{\Sigma}_3)_{[mnp} \partial_{q]} A = 3 (\tilde{\Sigma}_4)_{l_1 l_2 [mn} G^{l_1 l_2}{}_{pq]}$

$6 e^{-A} w_0 (\Sigma_7)_{mnpqrst} + 35 (\tilde{\Sigma}_3)_{[mnp} G_{qrst]} = 0$

Table A.2: Constraints from $\delta\Psi_\mu = 0$ coming from $\eta_+^T \gamma_{[n]} \text{Ext} - \text{Ext}^T \gamma_{[n]} \eta_+$.

$36 e^{-A} w_0 (\Sigma_1)_m + (\tilde{\Sigma}_3)^{l_1 l_2 l_3} G_{l_1 l_2 l_3 m} = 0$

$18 (\tilde{\Sigma}_3)_{l_1 mn} \partial^{l_1} A + (\tilde{\Sigma}_4)_{l_1 l_2 l_3 [m} G^{l_1 l_2 l_3}{}_{n]} = 18 e^{-A} w_0 (\Sigma_2)_{mn}$

$15 (\tilde{\Sigma}_4)_{[mnpq} \partial_{r]} A + 5 (\tilde{\Sigma}_3)_{l_1 [mn} G^{l_1}{}_{pqr]} = 3 e^{-A} w_0 (\Sigma_5)_{mnpqr}$

$3 e^{-A} w_0 (\Sigma_6)_{mnpqrs} + 10 (\tilde{\Sigma}_4)_{l_1 [mnp} G^{l_1}{}_{qrs]} = 0$

Table A.3: Constraints from $\delta\Psi_\mu = 0$ coming from $\eta_+^\dagger \gamma_{[n]} \text{Ext} + \text{Ext}^\dagger \gamma_{[n]} \eta_+$.

$144 (\Sigma_1)_{l_1} \partial^{l_1} A + (\Sigma_4)_{l_1 l_2 l_3 l_4} G^{l_1 l_2 l_3 l_4} = 0$

$144 \Sigma_0 \partial_m A + (\Sigma_5)_{m l_1 l_2 l_3 l_4} G^{l_1 l_2 l_3 l_4} = 0$

$144 (\Sigma_3)_{mn l_1} \partial^{l_1} A + (\Sigma_6)_{mn l_1 l_2 l_3 l_4} G^{l_1 l_2 l_3 l_4} = 12 G_{mn l_1 l_2} (\Sigma_2)^{l_1 l_2}$

$2 e^{-A} w_0^* (\tilde{\Sigma}_3^*)_{mnp} - 2 e^{-A} w_0 (\tilde{\Sigma}_3)_{mnp} + 12 (\Sigma_2)_{[mn} \partial_{p]} A + \frac{1}{36} (\Sigma_7)_{mnp l_1 l_2 l_3 l_4} G^{l_1 l_2 l_3 l_4}$
$= (\Sigma_3)^{l_1 l_2}{}_{[m} G_{np] l_1 l_2}$

$e^{-A} w_0 (\tilde{\Sigma}_4^*)_{mnpq} + e^{-A} w_0^* (\tilde{\Sigma}_4)_{mnpq} + 2 (\Sigma_5)_{l_1 mnpq} \partial^{l_1} A + \frac{1}{3} G_{mnpq} \Sigma_0$
$= 1 (\Sigma_4)_{l_1 l_2 [mn} G^{l_1 l_2}{}_{pq]}$

$6 (\Sigma_4)_{[mnpq} \partial_{r]} A + (\Sigma_1)_{[m} G_{npqr]} = (\Sigma_5)_{l_1 l_2 [mnp} G^{l_1 l_2}{}_{qr]}$

$5 (\Sigma_6)_{l_1 l_2 [mnpq} G^{l_1 l_2}{}_{rs]} = 4 (\Sigma_7)_{l_1 mnpqrs} \partial^{l_1} A + 10 (\Sigma_2)_{[mn} G_{pqrs]}$

$12 (\Sigma_6)_{[mnpqrs} \partial_{t]} A + 10 (\Sigma_3)_{[mnp} G_{qrst]} = 3 (\Sigma_7)_{l_1 l_2 [mnpqr} G^{l_1 l_2}{}_{st]}$

Table A.4: Constraints from $\delta\Psi_\mu = 0$ coming from $\eta_+^\dagger \gamma_{[n]}\text{Ext-Ext}^\dagger \gamma_{[n]}\eta_+$.

$\mu\Sigma_0 = 0$

$\frac{4}{3}i\mu(\Sigma_1)_m = 4(\Sigma_2)_{l_1 m}\partial^{l_1}A + \frac{1}{9}(\Sigma_3)^{l_1 l_2 l_3}G_{l_1 l_2 l_3 m}$

$6i\mu(\Sigma_2)_{mn} + 36(\Sigma_1)_{[m}\partial_{n]}A + (\Sigma_4)_{l_1 l_2 l_3 [m}G^{l_1 l_2 l_3}{}_{n]} = 0$

$4i\mu(\Sigma_3)_{mnp} + 2(\Sigma_1)^{l_1}G_{l_1 mnp}$
$= 6e^{-A}w_0(\tilde{\Sigma}_3^*)_{mnp} + 6e^{-A}w_0^*(\tilde{\Sigma}_3)_{mnp} + 12(\Sigma_4)_{l_1 mnp}\partial^{l_1}A + (\Sigma_5)_{l_1 l_2 l_3 [mn}G^{l_1 l_2 l_3}{}_{p]}$

$9e^{-A}w_0(\tilde{\Sigma}_4^*)_{mnpq} + 6i\mu(\Sigma_4)_{mnpq} + 72(\Sigma_3)_{[mnp}\partial_{q]}A + 2(\Sigma_6)_{l_1 l_2 l_3 [mnp}G^{l_1 l_2 l_3}{}_{q]}$
$= 9e^{-A}w_0^*(\tilde{\Sigma}_4)_{mnpq} + 12(\Sigma_2)_{l_1[m}G^{l_1}{}_{npq]}$

$12i\mu(\Sigma_5)_{mnpqr} + 60(\Sigma_3)_{l_1[mn}G^{l_1}{}_{pqr]}$
$= 36(\Sigma_6)_{l_1 mnpqr}\partial^{l_1}A + 5(\Sigma_7)_{l_1 l_2 l_3 [mnp}G^{l_1 l_2 l_3}{}_{r]}$

$i\mu(\Sigma_6)_{mnpqrs} + 18(\Sigma_5)_{[mnpqr}\partial_{s]}A = 10(\Sigma_4)_{l_1[mnp}G^{l_1}{}_{qrs]}$

$2i\mu(\Sigma_7)_{mnpqrst} + 35(\Sigma_5)_{l_1[mnpq}G^{l_1}{}_{rst]} = 0$

A.6 SUSY conditions for eleven-dimensional SUGRA without boundary

As we have mentioned in section 5.2 there are two ways to decompose the Majorana-spinor ϵ of eleven-dimensional supergravity such that one obtains $\mathcal{N} = 1$ SUSY compactifications on SU(3) structure manifolds. One of these possibilities was associated to the case in which boundaries are present, and which gives rise to the heterotic theory after dimensional reduction. The other option has to be used if one wants to reach type IIA string theory in ten dimensions, since only for this theory it is possible to have two internal spinors of opposite chirality. We discuss in this appendix what results can be obtained for the supersymmetry conditions in this case, and find agreement with [208], where the same problem has been studied. Moreover, we give a classification in terms of torsion classes and compare to the type IIA results after dimensional reduction, which has, up to our knowledge, not been done in the literature so far.

Spinor decomposition

We begin by decomposing the Majorana spinor ϵ in an appropriate way in order to give a non-chiral theory in the ten-dimensional limit. Along the lines of [208] the decomposition is given by

$$\epsilon = \chi_+ \otimes \xi + \chi_- \otimes \xi^* = \chi_+ \otimes (a'\,\eta'_+ + b'\,\eta'_-) + \text{c.c.}\,, \qquad (A.6.1)$$

with $a' \neq b'$. The two spinors that appear in the reduction to type IIA string theory would then be

$$\eta_1 = a' \eta'_+, \qquad \eta_2 = b' \eta'_-, \qquad (A.6.2)$$

where η_1 has positive chirality and η_2 has negative chirality, respectively. The two spinors η'_+ and η'_- can be used to define an invariant one-form $v = v_m dx^m$ and a seven-dimensional Majorana spinor η' of unit length

$$\eta'_+ = \frac{1}{\sqrt{2}} e^{\frac{Z'}{2}} (1 + v_m \gamma^m) \eta', \qquad (\eta'_+)^* = \eta'_- = \frac{1}{\sqrt{2}} e^{\frac{Z'}{2}} (1 - v_m \gamma^m) \eta'. \qquad (A.6.3)$$

The scalars $a' = e^{i\alpha'} |a'|$ and $b' = e^{i\beta'} |b'|$, appearing in (A.6.1) are in general complex functions of the internal space and could be used to absorb the factors of $e^{Z'/2}$ appearing in the definition of η' such that also η'_+ is of unit length $\|\eta'_+\|^2 = 1$. On the other hand, by defining $\eta_+ = a' \eta'_+$ and $b = b'/(a')^*$ one obtains a different form of (A.6.1)

$$\epsilon = \chi_+ \otimes \xi + \chi_- \otimes \xi^* = \chi_+ \otimes (\eta_+ + b\eta_-) + c.c., \qquad (A.6.4)$$

in which the spinor η_+ appears that we used to construct the SU(3) structure forms in section 3.3 and chapter 5. The Majorana spinor η' is then related to the G_2 structure spinor η of section 3.3 by

$$\eta' = (\cos \alpha' - i \sin \alpha' \, v_m \gamma^m) \eta. \qquad (A.6.5)$$

This is consistent with the fact that one can express η_+ in terms of the function e^Z, the form v, and the spinor η

$$\eta_+ = \frac{|a'|}{\sqrt{2}} e^{(\frac{Z'}{2} + i\alpha')} (1 + v_m \gamma^m) \eta' = \frac{1}{\sqrt{2}} e^{\frac{Z}{2}} (1 + v_m \gamma^m) \eta. \qquad (A.6.6)$$

Demanding $\|\eta'_+\|^2 = 1$ leads then to $\|\eta_+\|^2 = e^Z = |a'|^2$ as in section 3.3.

Then, there are two different sets of p-forms that one can define. On the one hand, there are the forms (3.3.11) constructed from η_+. On the other hand, one can build p-forms in the same way with the spinor ξ

$$\Xi_p = \xi^\dagger \gamma_{n_1 \ldots n_p} \xi \, dx^{n_1 \ldots n_p}, \qquad \tilde{\Xi}_p = \xi^T \gamma_{n_1 \ldots n_p} \xi \, dx^{n_1 \ldots n_p}. \qquad (A.6.7)$$

Due to (A.6.4) one can connect the Σ-forms to the Ξ-forms by

$$\begin{aligned}\Xi_p &= \Sigma_p + (-1)^p |b|^2 \Sigma_p^* + b^* \tilde{\Sigma}_p + (-1)^p b \tilde{\Sigma}_p^*, \\ \tilde{\Xi}_p &= \tilde{\Sigma}_p + (-1)^p b^2 \tilde{\Sigma}_p^* + b \Sigma_p + (-1)^p b \Sigma_p^*.\end{aligned} \quad (A.6.8)$$

Inverting these equations one finds

$$\begin{aligned}\Sigma_p &= \frac{1}{(1-|b|^2)^2} \left\{ \Xi_p + (-1)^p |b|^2 \Xi_p^* - b^* \tilde{\Xi}_p - (-1)^p b \tilde{\Xi}_p^* \right\}, \\ \tilde{\Sigma}_p &= \frac{1}{(1-|b|^2)^2} \left\{ \tilde{\Xi}_p + (-1)^p b^2 \tilde{\Xi}_p^* - b \Xi_p - (-1)^p b \Xi_p^* \right\},\end{aligned} \quad (A.6.9)$$

which will be useful later when we examine the SUSY conditions. However, one should note that (A.6.9) is only valid for $|b|^2 \neq 1$. Therefore, one has to do a case-by-case analysis. We will start by considering the case $|b|^2 = 1$.

The case $|b|^2 = 1$

For b being a pure phase, $b = e^{i\beta}$, the forms Ξ_1, Ξ_2, Ξ_5, Ξ_6, $\tilde{\Xi}_1$, $\tilde{\Xi}_2$, $\tilde{\Xi}_5$, and $\tilde{\Xi}_6$ are zero, due to the fact that the Σ_p's are purely real or purely imaginary, respectively. Furthermore, each of the remaining Ξ_p's is related to $\tilde{\Xi}_p$

$$\begin{aligned}\Xi_0 &= e^{-i\beta} \tilde{\Xi}_0 = 2\Sigma_0, \\ \Xi_3 &= e^{-i\beta} \tilde{\Xi}_3 = 2\Sigma_3 + 2i \operatorname{Im}(e^{-i\beta} \tilde{\Sigma}_3), \\ \Xi_4 &= e^{-i\beta} \tilde{\Xi}_4 = 2\Sigma_4 + 2 \operatorname{Re}(e^{-i\beta} \tilde{\Sigma}_4), \\ \Xi_7 &= e^{-i\beta} \tilde{\Xi}_7 = 2\Sigma_7.\end{aligned} \quad (A.6.10)$$

This reminds one of the case of G_2 structure, where also only the three-form, the four-form, and the volume-form are non-zero, as is discussed in section 3.3. We will now show that for a supersymmetric setting the case $|b|^2 = 1$ reduces to a G_2 structure.

Supersymmetry conditions

Plugging the compactification ansatz (2.4.1) – (2.4.3) into the SUSY variation of the gravitino (2.1.4) gives together with the spinor decomposition (A.6.1) the two equations

$$\delta\Psi_\mu = 0 \;\Rightarrow\; \text{Ext} \;=\; e^{-A} w_0 \, \xi^* + \left(\partial\!\!\!/ A + \frac{1}{6} G\!\!\!/ + \frac{i\mu}{3}\right) \xi \;=\; 0, \tag{A.6.11a}$$

$$\delta\Psi_m = 0 \;\Rightarrow\; \nabla_m \xi \;=\; \frac{1}{288} \left(i\mu\gamma_m + 8 G_{mpqr}\gamma^{pqr} - G_{npqr}\gamma_m{}^{npqr}\right)\xi, \tag{A.6.11b}$$

where w_0 is related to the AdS$_4$ radius by $R = 1/|w_0|$ and comes from the AdS$_4$ Killing spinor equation $\nabla_\mu \chi_+ = 1/2 \, \bar{w}_0 \, \hat{\gamma}_\mu \chi_-$, while $w_0 = 0$ for Minkowski space time. The first of these conditions will yield algebraic constraints on the flux, while the internal components will give differential constraints on Ξ_p and $\tilde{\Xi}_p$.

Combining the differential constraint on Ξ_0 and $\tilde{\Xi}_0$ gives

$$\mathrm{d}\Xi_0 \;=\; 0 \;=\; \mathrm{d}(e^{-i\beta}\tilde{\Xi}_0) \;=\; e^{-i\beta}\left(-i\,\mathrm{d}\beta\,\tilde{\Xi}_0 + \mathrm{d}\tilde{\Xi}_0\right) \;=\; -i\,\mathrm{d}\beta\,\Xi_0 - 2\,\mathrm{d}A\,\Xi_0. \tag{A.6.12}$$

Since both, $\mathrm{d}\beta$ and $\mathrm{d}A$ are real functions, respectively, this can only be true if $\mathrm{d}\beta = 0$ as well as $\mathrm{d}A = 0$. But for constant β one can make a gauge transformation to bring ϵ to the form

$$\epsilon \;=\; \chi_+ \otimes e^{\frac{i\beta}{2}}(\eta_+ + \eta_-) + c.c. \;=\; (e^{\frac{i\beta}{2}}\chi_+ + e^{-\frac{i\beta}{2}}\chi_-) \otimes \eta \;=\; \tilde{\chi}\otimes\eta, \tag{A.6.13}$$

where $\tilde{\chi}$ as well as η are Majorana spinors. So we see that by SUSY the SU(3) ansatz (A.6.1) with $|a'|^2 = |b'|^2$ (which is equivalent to $|b|^2 = 1$) leads to the G$_2$ structure ansatz (A.6.13) and is therefore excluded from an SU(3) structure analysis. We hence turn to the case $|b| \neq 1$.

The case $|b| \neq 1$

In order to analyze the case $|b| \neq 1$, we use again the compactification ansatz (2.4.1) – (2.4.3). The spinor decomposition A.6.4 leads again to the SUSY conditions (A.6.11a) and (A.6.11b). But now Ξ_p and $\tilde{\Xi}_p$ do not take a form as easy as in the previous section, which makes the analysis much more complicated.

For $|b|^2 \neq 1$ we can use equation (A.6.9) in order to express Σ_p and $\tilde{\Sigma}_p$ in terms of Ξ_p and $\tilde{\Xi}_p$, respectively. This is an advantage, since we are in the end not interested in the exterior derivatives of Ξ_p, which we can get easily from (A.6.11b), but in the derivatives of the SU(3) structure forms in seven dimensions v, J, and Ω which are given in terms of Σ's by

$$v \;=\; e^{-Z}\Sigma_1, \qquad J \;=\; i\,e^{-Z}\Sigma_2, \qquad \Omega \;=\; i\,e^{-Z}\tilde{\Sigma}_3. \tag{A.6.14}$$

Again, v is perpendicular to J and Ω, and thus M looks locally like the direct product of an SU(3) structure manifold and a line, while globally v gives the direction of the S^1 that shrinks to zero by going from M-theory to type IIA string theory.

Constraints for the flux

Before we consider the differential constraints from (A.6.11b) in more detail, we examine which general conditions on the flux can be gained from (A.6.11a). Acting on this equation with $\xi^\dagger$ and ξ^T times n antisymmetrized gamma matrices yields the following restrictions on G_{11}

$$dA = \frac{1}{12}\frac{1-|b|^2}{1+|b|^2}G\lrcorner(v\wedge J^2) \tag{A.6.15}$$

$$= \frac{1}{2}(1-|b|^2)e^{-A}\mathrm{Re}\left(\frac{w_0}{b}\right)v + \frac{1}{12}(1-|b|^2)G\lrcorner\mathrm{Im}\left(\frac{\Omega}{b}\right),$$

$$e^{-A}\mathrm{Im}\left(\frac{w_0}{b}\right) = -\frac{1+|b|^2}{6|b|^2}\mu \tag{A.6.16}$$

$$= -\frac{1}{6}\frac{1+|b|^2}{1-|b|^2}G\lrcorner\left(v\wedge\mathrm{Re}\left(\frac{\Omega}{b}\right)\right),$$

$$H\lrcorner J = \frac{1}{2}(1+|b|^2)F\lrcorner\mathrm{Re}\left(\frac{\Omega}{b}\right), \tag{A.6.17}$$

where we used $G = F - v \wedge H$. Using the decomposition (5.2.6)

$$\begin{aligned} F &= A_1 J \wedge J + A_2 \wedge J + \overline{B}\wedge\Omega + B\wedge\overline{\Omega}, \\ H &= \overline{C}_1\Omega + C_1\overline{\Omega} + C_2\wedge J + C_3, \end{aligned} \tag{A.6.18}$$

one finds

$$dA\lrcorner v = (1-|b|^2)\left[\frac{1}{2}e^{-A}\mathrm{Re}\left(\frac{w_0}{b}\right) + \frac{2}{3}\mathrm{Im}\left(\frac{C_1}{b}\right)\right] \tag{A.6.19}$$

$$= \frac{1-|b|^2}{1+|b|^2}A_1,$$

$$dA_\perp = -\frac{2}{3}(1-|b|^2)\mathrm{Im}\left(\frac{B}{b}\right) \tag{A.6.20}$$

$$= -\frac{1}{3}\frac{1-|b|^2}{1+|b|^2}C_2\lrcorner J,$$

$$e^{-A} \operatorname{Im}\left(\frac{w_0}{b}\right) = -\frac{1+|b|^2}{6|b|^2}\mu \qquad (A.6.21)$$
$$= \frac{4}{3}\frac{1+|b|^2}{1-|b|^2}\operatorname{Re}\left(\frac{C_1}{b}\right).$$

So we see that dA and w_0/b are fixed by the flux.

Differential constraints

The differential constraints for v, J, and Ω can be obtained in the following way. First, calculate the exterior derivatives of Ξ_p and $\tilde{\Xi}_p$ and use (A.6.8) to transform them into expressions depending on v, J, and Ω. Then, use (A.6.9) to express the exterior derivatives of Σ_p or $\tilde{\Sigma}_p$ in terms of $d\Xi_p$ and $d\tilde{\Xi}_p$. Since this is quite laborious, we will only give the main results. A special case is given by the derivative of $\Sigma_0 = e^Z$. It can be obtained either by starting from Ξ_0 or from $\tilde{\Xi}_0$, and gives thus a consistency condition which reads

$$dZ = -d\ln(1+|b|^2) + dA = -d\ln b - 4\, dA + (1-|b|^2)\, e^{-A}\frac{w_0}{b}\frac{\Sigma_1}{\Sigma_0}. \qquad (A.6.22)$$

Decomposing this equation into real and imaginary part gives one equation for the absolute value and for the argument of b, respectively

$$d|b|^2 = 2(1+|b|^2)\,|b|^2\, e^{-A}\operatorname{Re}\left(\frac{w_0}{b}\right)v - 6|b|^2\frac{1+|b|^2}{1-|b|^2}\, dA, \qquad (A.6.23)$$
$$d\beta = -(1-|b|^2)\, e^{-A}\operatorname{Im}\left(\frac{w_0}{b}\right)v.$$

These equations can be used to eliminate $d|b|^2$ and $d\beta$ in the exterior derivatives of SU(3) structure forms. After a tedious calculation one finds

$$dv = -2\frac{1+4|b|^2+|b|^4}{(1-|b|^2)^2}\, dA \wedge v, \qquad (A.6.24)$$

$$dJ = -\left[4\frac{1+|b|^2+|b|^4}{(1-|b|^2)^2}\, dA + 2\frac{|b|^2}{1-|b|^2}\, e^{-A}\operatorname{Re}\left(\frac{w_0}{b}\right)v\right]\wedge J \qquad (A.6.25)$$
$$-3|b|^2\frac{1+|b|^2}{1-|b|^2}\, e^{-A}\operatorname{Re}\left(\frac{w_0}{b}\right)\operatorname{Re}\left(\frac{\Omega}{b}\right)$$
$$-3|b|^2\, e^{-A}\operatorname{Im}\left(\frac{w_0}{b}\right)\operatorname{Im}\left(\frac{\Omega}{b}\right),$$
$$+\frac{1+|b|^2}{1-|b|^2}*G$$

$$\begin{aligned}
\frac{\mathrm{d}\Omega}{b} &= -\left[3\,\mathrm{d}A + 2i\,e^{-A}\mathrm{Im}\left(\frac{w_0}{b}\right)v + 6\frac{|b|^2}{1-|b|^2}e^{-A}\mathrm{Re}\left(\frac{w_0}{b}\right)v\right]\wedge\mathrm{Re}\left(\frac{\Omega}{b}\right) \\
&\quad -\left[3i\frac{(1+|b|^2)^2}{(1-|b|^2)^2}\mathrm{d}A + 2|b|^2\frac{2-|b|^2}{1+|b|^2}\mathrm{Im}\left(\frac{w_0}{b}\right)v\right]\wedge\mathrm{Im}\left(\frac{\Omega}{b}\right) \quad\text{(A.6.26)}\\
&\quad +\left[e^{-A}\mathrm{Im}\left(\frac{w_0}{b}\right) - i\frac{1+|b|^2}{1-|b|^2}e^{-A}\mathrm{Re}\left(\frac{w_0}{b}\right)\right]J^2 \\
&\quad + 12\frac{1+|b|^2}{(1-|b|^2)^2}\mathrm{d}A\wedge v\wedge J \\
&\quad + \frac{2}{1-|b|^2}v\wedge *G - \frac{2i}{1-|b|^2}G\,, \\
\mathrm{d}J^2 &= -\left[2\mathrm{d}A + 4\frac{|b|^2}{1-|b|^2}e^{-A}\mathrm{Re}\left(\frac{w_0}{b}\right)\right]\wedge J^2 \quad\text{(A.6.27)}\\
&\quad + 24|b|^2\frac{1+|b|^2}{(1-|b|^2)^2}\mathrm{d}A\wedge v\wedge\mathrm{Im}\left(\frac{\Omega}{b}\right) \\
&\quad - 2\frac{1+|b|^2}{1-|b|^2}v\wedge G\,.
\end{aligned}$$

A further simplification can be obtained by demanding that the exterior derivative of (A.6.23) has to be zero. Together with the fact that w_0 is a constant this gives a consistency condition on $\mathrm{d}|b|^2$ that can only be satisfied when $\mathrm{d}A_\perp = 0$. This implies that $\mathrm{Im}(B/b) = 0$, but since B is a $(1,0)$-form this can only be true for $B = 0$. Analogously one finds that $C_2 = 0$.

Torsion classes

One can also transform the constraints given in the last two sections into constraints on the torsion classes of the manifold (3.3.17)

$$\begin{aligned}
\mathrm{d}v &= RJ + \bar{V}_1\lrcorner\Omega + V_1\lrcorner\bar{\Omega} + v\wedge W_0 + T_1\,, &\text{(A.6.28)}\\
\mathrm{d}J &= -\frac{3}{2}\mathrm{Im}(\bar{W}_1\,\Omega) + W_4\wedge J + W_3 + v\wedge(\frac{2}{3}\mathrm{Re}E\,J + \bar{V}_2\lrcorner\Omega + V_2\lrcorner\bar{\Omega} + T_2)\,, \\
\mathrm{d}\Omega &= W_1 J\wedge J + W_2\wedge J + \bar{W}_5\wedge\Omega + v\wedge(E\Omega - 4V_2\wedge J + S)\,.
\end{aligned}$$

For the scalar torison classes we find

$$
\begin{aligned}
R &= 0, \\
\operatorname{Re}\frac{W_1}{b} &= e^{-A}\operatorname{Im}\frac{w_0}{b}, \\
\operatorname{Im}\frac{W_1}{b} &= -2\frac{1}{1-|b|^2}A_1 - \frac{1+|b|^2}{1-|b|^2}e^{-A}\operatorname{Re}\frac{w_0}{b}, \\
\operatorname{Re}E &= -3\frac{1+|b|^4}{1-|b|^4}A_1 - 3\frac{|b|^2}{1-|b|^2}e^{-A}\operatorname{Re}\frac{w_0}{b}, \\
\operatorname{Im}E &= |b|^2\frac{1-2|b|^2}{1+|b|^2}e^{-A}\operatorname{Im}\frac{w_0}{b}, \\
\operatorname{Re}\frac{C_1}{b} &= \frac{3}{4}\frac{1-|b|^2}{1+|b|^2}e^{-A}\operatorname{Im}\frac{w_0}{b}, \\
\operatorname{Im}\frac{C_1}{b} &= \frac{3}{2}\frac{1}{1+|b|^2}A_1 - \frac{3}{4}e^{-A}\operatorname{Im}\frac{w_0}{b},
\end{aligned}
\qquad (A.6.29)
$$

while the vector torsion classes all have to vanish, due to $B = C_2 = 0$

$$
\begin{aligned}
V_1 &= 0, \\
V_2 &= i\frac{1+|b|^2}{1-|b|^2}B = 0, \\
W_0 &= \frac{4}{3}\frac{1+4|b|^2+|b|^4}{1-|b|^2}\operatorname{Im}\frac{B}{b} = 0, \\
W_4 &= \frac{2}{3}(1-|b|^2)\operatorname{Im}\frac{B}{b} = 0, \\
W_5 &= -i(1-|b|^2)\frac{B}{b} = 0, \\
B &= 0, \\
C_2 &= 0.
\end{aligned}
\qquad (A.6.30)
$$

The remaining torsion classes have to obey

$$
\begin{aligned}
W_2 &= -\frac{2ib}{1-|b|^2}A_2 = -\frac{2ib}{1+|b|^2}T_2 \\
S &= -\frac{2b}{1-|b|^2}\left(\frac{1-|b|^2}{1+|b|^2}W_3 - iC_3\right) = \frac{4ib}{1-|b|^2}C_3^{2,1}, \\
T_1 &= 0.
\end{aligned}
\qquad (A.6.31)
$$

Limits

We see that all scalar torsion classes depend on the $(2,2)$ component of G, A_1, and the AdS parameter w_0. Setting $w_0 = 0$, i.e. considering a compactification to warped Minkowski space time, allows only for non-zero $\mathrm{Im}(W_1/b)$, $\mathrm{Re}E$, and $\mathrm{Re}(C_1/b)$, which depend all only on A_1. Reducing the theory now to type IIA supergravity sets the torsion classes governing the fibration over v to zero. This means that $E = V_2 = T_2 = S = 0$. So we see that for a supersymmetric reduction to type II SUGRA on Minkowski spacetime all flux components and all torsion classes have to be zero. This is in accordance with [160]. There, it was shown that for type IIA compactifications to Minkowski space fluxes can only be present if $b = 0$ or if $|b| = 1$, which is excluded by our parametrization.

On the other hand, allowing for an AdS space-time a supersymmetric reduction to ten dimensions sets $\mathrm{Im}(w_0/b) = 0$ and relates A_1 and $\mathrm{Re}(w_0/b)$. This allows only for flux of the form $G = F^{2,2} + v \wedge (H^{3,0} + H^{0,3})$ and a non-vanishing torsion class W_1, which all depend only on $e^{-A}\mathrm{Re}(w_0/b)$.

Bibliography

[1] J. Held, D. Lüst, F. Marchesano, and L. Martucci, "DWSB in heterotic flux compactifications," *JHEP* **1006** (2010) 090, arXiv:1004.0867 [hep-th].

[2] J. Held, "DWSB for heterotic flux compactifications," *Fortsch.Phys.* **59** (2011) 711–715, arXiv:1103.2636 [hep-th].

[3] J. Held, "Heterotic domain-wall supersymmetry breaking," *Nucl.Phys.Proc.Suppl.* **216** (2011) 239–240.

[4] J. Held, "BPS-like potential for compactifications of heterotic M-theory?," *JHEP* **1110** (2011) 136, arXiv:1109.1974 [hep-th].

[5] M. Green, J. Schwarz, and E. Witten, *Superstring Theory: Introduction*. Cambridge Monographs on Mathematical Physics. Cambridge University Press, 1988.

[6] M. Green, J. Schwarz, and E. Witten, *Superstring Theory: Loop amplitudes, anomalies, and phenomenology*. Cambridge Monographs on Mathematical Physics. Cambridge University Press, 1987.

[7] D. Lüst and S. Theisen, *Lectures on string theory*. Lecture notes in physics. Springer-Verlag, 1989.

[8] J. Polchinski, *String Theory: An introduction to the Bosonic String*. No. Bd. 1 in Cambridge Monographs on Mathematical Physics. Cambridge University Press, 2005.

[9] J. Polchinski, *String Theory: Superstring theory and beyond*. No. Bd. 2 in Cambridge Monographs on Mathematical Physics. Cambridge University Press, 2005.

[10] K. Becker, M. Becker, and J. Schwarz, *String Theory And M-Theory: A Modern Introduction*. Cambridge University Press, 2007.

[11] J. M. Maldacena, "The Large N limit of superconformal field theories and supergravity," *Adv.Theor.Math.Phys.* **2** (1998) 231–252, arXiv:hep-th/9711200 [hep-th].

[12] C. Vafa, "Evidence for F theory," *Nucl.Phys.* **B469** (1996) 403–418, arXiv:hep-th/9602022 [hep-th].

[13] L. Susskind, "The Anthropic landscape of string theory," arXiv:hep-th/0302219 [hep-th].

[14] S. Kachru, R. Kallosh, A. D. Linde, J. M. Maldacena, L. P. McAllister, *et al.*, "Towards inflation in string theory," *JCAP* **0310** (2003) 013, arXiv:hep-th/0308055 [hep-th].

[15] A. Strominger and C. Vafa, "Microscopic origin of the Bekenstein-Hawking entropy," *Phys.Lett.* **B379** (1996) 99–104, arXiv:hep-th/9601029 [hep-th].

[16] C. Hull and B. Zwiebach, "Double Field Theory," *JHEP* **0909** (2009) 099, arXiv:0904.4664 [hep-th].

[17] O. Hohm, C. Hull, and B. Zwiebach, "Background independent action for double field theory," *JHEP* **1007** (2010) 016, arXiv:1003.5027 [hep-th].

[18] J. Shelton, W. Taylor, and B. Wecht, "Nongeometric flux compactifications," *JHEP* **0510** (2005) 085, arXiv:hep-th/0508133 [hep-th].

[19] A. Dabholkar and C. Hull, "Generalised T-duality and non-geometric backgrounds," *JHEP* **0605** (2006) 009, arXiv:hep-th/0512005 [hep-th].

[20] J. Scherk, "An Introduction to the Theory of Dual Models and Strings," *Rev.Mod.Phys.* **47** (1975) 123–164.

[21] J. Schwarz, *Superstrings. The first 15 years of superstring theory*. No. Bd. 1-2. Singapore, 1985.

[22] J. H. Schwarz, "The Early Years of String Theory: A Personal Perspective," arXiv:0708.1917 [hep-th]. Based on a lecture presented on June 20, 2007 at the Galileo Galilei Institute.

[23] F. Gliozzi, J. Scherk, and D. I. Olive, "Supersymmetry, Supergravity Theories and the Dual Spinor Model," *Nucl.Phys.* **B122** (1977) 253–290.

[24] M. B. Green and J. H. Schwarz, "Anomaly Cancellation in Supersymmetric D=10 Gauge Theory and Superstring Theory," *Phys.Lett.* **B149** (1984) 117–122.

[25] D. J. Gross, J. A. Harvey, E. J. Martinec, and R. Rohm, "Heterotic String Theory. 1. The Free Heterotic String," *Nucl.Phys.* **B256** (1985) 253.

[26] D. J. Gross, J. A. Harvey, E. J. Martinec, and R. Rohm, "Heterotic String Theory. 2. The Interacting Heterotic String," *Nucl.Phys.* **B267** (1986) 75.

[27] P. Candelas, G. T. Horowitz, A. Strominger, and E. Witten, "Vacuum Configurations for Superstrings," *Nucl.Phys.* **B258** (1985) 46–74.

[28] I. Antoniadis, N. Arkani-Hamed, S. Dimopoulos, and G. Dvali, "New dimensions at a millimeter to a Fermi and superstrings at a TeV," *Phys.Lett.* **B436** (1998) 257–263, arXiv:hep-ph/9804398 [hep-ph].

[29] N. Arkani-Hamed, S. Dimopoulos, and G. Dvali, "Phenomenology, astrophysics and cosmology of theories with submillimeter dimensions and TeV scale quantum gravity," *Phys.Rev.* **D59** (1999) 086004, arXiv:hep-ph/9807344 [hep-ph].

[30] D. Lüst, S. Stieberger, and T. R. Taylor, "The LHC String Hunter's Companion," *Nucl.Phys.* **B808** (2009) 1–52, arXiv:0807.3333 [hep-th].

[31] D. Lüst, O. Schlotterer, S. Stieberger, and T. Taylor, "The LHC String Hunter's Companion (II): Five-Particle Amplitudes and Universal Properties," *Nucl.Phys.* **B828** (2010) 139–200, arXiv:0908.0409 [hep-th].

[32] D. Lüst, "Seeing through the String Landscape - a String Hunter's Companion in Particle Physics and Cosmology," *JHEP* **0903** (2009) 149, arXiv:0904.4601 [hep-th].

[33] **CMS** Collaboration, V. Khachatryan *et al.*, "Search for Dijet Resonances in 7 TeV pp Collisions at CMS," *Phys.Rev.Lett.* **105** (2010) 211801, arXiv:1010.0203 [hep-ex].

[34] P. H. Ginsparg, "Applied conformal field theory," arXiv:hep-th/9108028 [hep-th].

[35] M. R. Gaberdiel, "An Introduction to conformal field theory," *Rept.Prog.Phys.* **63** (2000) 607–667, arXiv:hep-th/9910156 [hep-th].

[36] P. D. Francesco, P. Mathieu, and D. Sénéchal, *Conformal field theory*. Graduate texts in contemporary physics. Springer, 1997.

[37] e. Schottenloher, Martin, "A mathematical introduction to conformal field theory," *Lect.Notes Phys.* **759** (2008) 1–237.

[38] R. Blumenhagen and E. Plauschinn, *Introduction to Conformal Field Theory: With Applications to String Theory*. Lecture notes in physics. Springer, 2009.

[39] A. Bilal, "Lectures on Anomalies," arXiv:0802.0634 [hep-th].

[40] J. A. Harvey, "TASI 2003 lectures on anomalies," arXiv:hep-th/0509097 [hep-th].

[41] L. Alvarez-Gaume, D. Z. Freedman, and S. Mukhi, "The Background Field Method and the Ultraviolet Structure of the Supersymmetric Nonlinear Sigma Model," *Annals Phys.* **134** (1981) 85.

[42] D. H. Friedan, "Nonlinear Models in Two + Epsilon Dimensions," *Annals Phys.* **163** (1985) 318. Ph.D. Thesis.

[43] D. Friedan, E. J. Martinec, and S. H. Shenker, "Conformal Invariance, Supersymmetry and String Theory," *Nucl.Phys.* **B271** (1986) 93.

[44] J. Callan, Curtis G., E. Martinec, M. Perry, and D. Friedan, "Strings in Background Fields," *Nucl.Phys.* **B262** (1985) 593.

[45] M. B. Green and J. H. Schwarz, "Covariant Description of Superstrings," *Phys.Lett.* **B136** (1984) 367–370.

[46] M. B. Green and J. H. Schwarz, "Properties of the Covariant Formulation of Superstring Theories," *Nucl.Phys.* **B243** (1984) 285.

[47] M. B. Green and J. H. Schwarz, "Supersymmetrical Dual String Theory," *Nucl.Phys.* **B181** (1981) 502–530.

[48] M. B. Green and J. H. Schwarz, "Supersymmetrical Dual String Theory. 2. Vertices and Trees," *Nucl.Phys.* **B198** (1982) 252–268.

[49] J. Dai, R. Leigh, and J. Polchinski, "New Connections Between String Theories," *Mod.Phys.Lett.* **A4** (1989) 2073–2083.

[50] J. Polchinski, "Dirichlet Branes and Ramond-Ramond charges," *Phys.Rev.Lett.* **75** (1995) 4724–4727, arXiv:hep-th/9510017 [hep-th].

[51] J. Polchinski, S. Chaudhuri, and C. V. Johnson, "Notes on D-branes," arXiv:hep-th/9602052 [hep-th].

[52] K. Narain, M. Sarmadi, and E. Witten, "A Note on Toroidal Compactification of Heterotic String Theory," *Nucl.Phys.* **B279** (1987) 369.

[53] M. Dine, P. Y. Huet, and N. Seiberg, "Large and Small Radius in String Theory," *Nucl.Phys.* **B322** (1989) 301.

[54] A. Font, L. E. Ibanez, D. Lüst, and F. Quevedo, "Supersymmetry breaking from duality invariant gaugino condensation," *Phys. Lett.* **B245** (1990) 401–408.

[55] J. Polchinski and E. Witten, "Evidence for heterotic - type I string duality," *Nucl.Phys.* **B460** (1996) 525–540, arXiv:hep-th/9510169 [hep-th].

[56] E. Witten, "String theory dynamics in various dimensions," *Nucl.Phys.* **B443** (1995) 85–126, arXiv:hep-th/9503124 [hep-th].

[57] P. Horava and E. Witten, "Eleven-dimensional supergravity on a manifold with boundary," *Nucl.Phys.* **B475** (1996) 94–114, arXiv:hep-th/9603142 [hep-th].

[58] P. Townsend, "Four lectures on M theory," arXiv:hep-th/9612121 [hep-th]. 55pp. 4 figs. Latex. Sprocl.sty macro needed. To appear in procs. of the ICTP summer school on High Energy Physics and Cosmology, Trieste, June 1996. Revision includes minor corrections to some formulae and addition of a reference.

[59] K. Kikkawa and M. Yamasaki, "Casimir Effects in Superstring Theories," *Phys.Lett.* **B149** (1984) 357.

[60] T. Buscher, "A Symmetry of the String Background Field Equations," *Phys.Lett.* **B194** (1987) 59.

[61] T. Buscher, "Path Integral Derivation of Quantum Duality in Nonlinear Sigma Models," *Phys.Lett.* **B201** (1988) 466.

[62] A. Giveon, M. Porrati, and E. Rabinovici, "Target space duality in string theory," *Phys.Rept.* **244** (1994) 77–202, arXiv:hep-th/9401139 [hep-th].

[63] T. Kaluza, "On the Problem of Unity in Physics," *Sitzungsber.Preuss.Akad.Wiss.Berlin (Math.Phys.)* **1921** (1921) 966–972. Often incorrectly cited as Sitzungsber.Preuss.Akad.Wiss.Berlin (Math. Phys.) K1,966. In reality there is no volume number, so SPIRES used the year in place of a volume number.

[64] O. Klein, "Quantum Theory and Five-Dimensional Theory of Relativity. (In German and English)," *Z.Phys.* **37** (1926) 895–906.

[65] J. Callan, Curtis G., J. A. Harvey, and A. Strominger, "Supersymmetric string solitons," arXiv:hep-th/9112030 [hep-th].

[66] A. Font, L. E. Ibanez, D. Lüst, and F. Quevedo, "Strong - weak coupling duality and nonperturbative effects in string theory," *Phys. Lett.* **B249** (1990) 35–43.

[67] J. H. Schwarz and A. Sen, "Duality symmetries of 4-D heterotic strings," *Phys. Lett.* **B312** (1993) 105–114, arXiv:hep-th/9305185.

[68] C. Hull and P. Townsend, "Unity of superstring dualities," *Nucl.Phys.* **B438** (1995) 109–137, arXiv:hep-th/9410167 [hep-th].

[69] P. Townsend, "The eleven-dimensional supermembrane revisited," *Phys.Lett.* **B350** (1995) 184–187, arXiv:hep-th/9501068 [hep-th].

[70] D. B. Kaplan, "Effective field theories," arXiv:nucl-th/9506035 [nucl-th].

[71] I. Z. Rothstein, "TASI lectures on effective field theories," arXiv:hep-ph/0308266 [hep-ph].

[72] A. V. Manohar, "Effective field theories," arXiv:hep-ph/9606222 [hep-ph].

[73] W. D. Goldberger, "Les Houches lectures on effective field theories and gravitational radiation," arXiv:hep-ph/0701129 [hep-ph].

[74] E. D'Hoker and D. Phong, "The Geometry of String Perturbation Theory," *Rev.Mod.Phys.* **60** (1988) 917.

[75] Z. Bern, L. J. Dixon, D. Dunbar, M. Perelstein, and J. Rozowsky, "On the relationship between Yang-Mills theory and gravity and its implication for ultraviolet divergences," *Nucl.Phys.* **B530** (1998) 401–456, arXiv:hep-th/9802162 [hep-th].

[76] L. F. Alday and J. M. Maldacena, "Gluon scattering amplitudes at strong coupling," *JHEP* **0706** (2007) 064, arXiv:0705.0303 [hep-th].

[77] E. D'Hoker and D. Phong, "Lectures on two loop superstrings," *Conf.Proc.* **C0208124** (2002) 85–123, arXiv:hep-th/0211111 [hep-th].

[78] D. Lüst, P. Mayr, R. Richter, and S. Stieberger, "Scattering of gauge, matter, and moduli fields from intersecting branes," *Nucl.Phys.* **B696** (2004) 205–250, arXiv:hep-th/0404134 [hep-th].

[79] W. Staessens and B. Vercnocke, "Lectures on Scattering Amplitudes in String Theory," arXiv:1011.0456 [hep-th].

[80] D. J. Gross and J. H. Sloan, "The Quartic Effective Action for the Heterotic String," *Nucl.Phys.* **B291** (1987) 41.

[81] R. Metsaev and A. A. Tseytlin, "Order alpha-prime (Two Loop) Equivalence of the String Equations of Motion and the Sigma Model Weyl Invariance Conditions: Dependence on the Dilaton and the Antisymmetric Tensor," *Nucl.Phys.* **B293** (1987) 385.

[82] C. Hull and P. Townsend, "The two loop beta function for sigma models with torsion," *Phys.Lett.* **B191** (1987) 115. Revised version.

[83] W. Nahm, "Supersymmetries and their Representations," *Nucl.Phys.* **B135** (1978) 149.

[84] P. Van Nieuwenhuizen, "Supergravity," *Phys.Rept.* **68** (1981) 189–398.

[85] A. H. Chamseddine, "Interacting Supergravity in Ten-Dimensions: The Role of the Six-Index Gauge Field," *Phys.Rev.* **D24** (1981) 3065.

[86] E. Bergshoeff, M. de Roo, B. de Wit, and P. van Nieuwenhuizen, "Ten-Dimensional Maxwell-Einstein Supergravity, Its Currents, and the Issue of Its Auxiliary Fields," *Nucl.Phys.* **B195** (1982) 97–136.

[87] G. Chapline and N. Manton, "Unification of Yang-Mills Theory and Supergravity in Ten-Dimensions," *Phys.Lett.* **B120** (1983) 105–109.

[88] M. B. Green and J. H. Schwarz, "Extended Supergravity in Ten-Dimensions," *Phys.Lett.* **B122** (1983) 143.

[89] E. Cremmer, B. Julia, and J. Scherk, "Supergravity Theory in Eleven-Dimensions," *Phys.Lett.* **B76** (1978) 409–412.

[90] P. Horava and E. Witten, "Heterotic and type I string dynamics from eleven-dimensions," *Nucl.Phys.* **B460** (1996) 506–524, `arXiv:hep-th/9510209 [hep-th]`.

[91] E. Witten, "Strong coupling expansion of Calabi-Yau compactification," *Nucl.Phys.* **B471** (1996) 135–158, `arXiv:hep-th/9602070 [hep-th]`.

[92] A. Lukas, B. A. Ovrut, and D. Waldram, "The Ten-dimensional effective action of strongly coupled heterotic string theory," *Nucl.Phys.* **B540** (1999) 230–246, `arXiv:hep-th/9801087 [hep-th]`.

[93] T. Banks, W. Fischler, S. Shenker, and L. Susskind, "M theory as a matrix model: A Conjecture," *Phys.Rev.* **D55** (1997) 5112–5128, `arXiv:hep-th/9610043 [hep-th]`.

[94] N. Seiberg, "Why is the matrix model correct?," *Phys.Rev.Lett.* **79** (1997) 3577–3580, `arXiv:hep-th/9710009 [hep-th]`.

[95] L. Randall and R. Sundrum, "An Alternative to compactification," *Phys.Rev.Lett.* **83** (1999) 4690–4693, `arXiv:hep-th/9906064 [hep-th]`.

[96] R. Blumenhagen, B. Kors, D. Lüst, and S. Stieberger, "Four-dimensional String Compactifications with D-Branes, Orientifolds and Fluxes," *Phys.Rept.* **445** (2007) 1–193, `arXiv:hep-th/0610327 [hep-th]`.

[97] C. Burgess and G. Moore, *The Standard Model: A Primer*. Cambridge University Press, 2007.

[98] S. P. Martin, "A Supersymmetry primer," `arXiv:hep-ph/9709356 [hep-ph]`.

[99] J. Wess and J. Bagger, *Supersymmetry and supergravity*. Princeton series in physics. Princeton University Press, 1992.

[100] I. Buchbinder and S. Kuzenko, *Ideas and methods of supersymmetry and supergravity, or, A walk through superspace*. Studies in high energy physics, cosmology, and gravitation. Institute of Physics Pub., 1998.

[101] L. Abbott, "The Background Field Method Beyond One Loop," *Nucl.Phys.* **B185** (1981) 189.

[102] J. Callan, Curtis G., C. Lovelace, C. Nappi, and S. Yost, "String Loop Corrections to beta Functions," *Nucl.Phys.* **B288** (1987) 525.

[103] J. Derendinger, S. Ferrara, C. Kounnas, and F. Zwirner, "On loop corrections to string effective field theories: Field dependent gauge couplings and sigma model anomalies," *Nucl.Phys.* **B372** (1992) 145–188. Revised version.

[104] B. de Wit, V. Kaplunovsky, J. Louis, and D. Lüst, "Perturbative couplings of vector multiplets in N=2 heterotic string vacua," *Nucl.Phys.* **B451** (1995) 53–95, `arXiv:hep-th/9504006 [hep-th]`.

[105] L. Alvarez-Gaume and E. Witten, "Gravitational Anomalies," *Nucl.Phys.* **B234** (1984) 269.

[106] A. Strominger and E. Witten, "New Manifolds for Superstring Compactification," *Commun.Math.Phys.* **101** (1985) 341.

[107] B. R. Greene, K. H. Kirklin, P. J. Miron, and G. G. Ross, "A Three Generation Superstring Model. 1. Compactification and Discrete Symmetries," *Nucl.Phys.* **B278** (1986) 667.

[108] B. R. Greene, K. H. Kirklin, P. J. Miron, and G. G. Ross, "A Three Generation Superstring Model. 2. Symmetry Breaking and the Low-Energy Theory," *Nucl.Phys.* **B292** (1987) 606.

[109] D. Gepner, "Exactly Solvable String Compactifications on Manifolds of SU(N) Holonomy," *Phys.Lett.* **B199** (1987) 380–388.

[110] E. Calabi, "On Kähler manifolds with vanishing canonical class," in *Algebraic Geometry and Topology: A Symposium in Honor of S. Lefschetz*, p. 78. Princeton University Press, 1955.

[111] S.-T. Yau, "Calabi's Conjecture and some new results in algebraic geometry," *Proc.Nat.Acad.Sci.* **74** (1977) 1798–1799.

[112] P. Candelas, "Lectures on complex manifolds," in *Superstrings '87*, pp. 1–88. Singapore: World Scientific, 1987.

[113] T. Hübsch, *Calabi-Yau manifolds: a bestiary for physicists*. World Scientific, 1994.

[114] K. Hori *et al.*, "Mirror symmetry," in *Clay Mathematical Monographs*, vol. 1. Providence: American Mathematical Society. Providence, USA: AMS (2003) 929 p.

[115] M. Awada, M. Duff, and C. Pope, "N=8 Supergravity Breaks Down to N=1," *Phys.Rev.Lett.* **50** (1983) 294.

[116] M. Duff, B. Nilsson, and C. Pope, "Compactification of d = 11 supergravity on K(3) x U(3)," *Phys.Lett.* **B129** (1983) 39.

[117] P. Candelas and D. Raine, "Spontaneous compactification and supersymmetry in d = 11 supergravity," *Nucl.Phys.* **B248** (1984) 415.

[118] E. Fischbach and C. Talmadge, "Six years of the fifth force," *Nature* **356** (1992) 207–214.

[119] S. Cecotti, S. Ferrara, and L. Girardello, "Geometry of Type II Superstrings and the Moduli of Superconformal Field Theories," *Int.J.Mod.Phys.* **A4** (1989) 2475.

[120] S. Ferrara and S. Sabharwal, "Dimensional reduction of type II superstrings," *Class.Quant.Grav.* **6** (1989) L77.

[121] S. Ferrara and S. Sabharwal, "Quaternionic Manifolds for Type II Superstring Vacua of Calabi-Yau Spaces," *Nucl.Phys.* **B332** (1990) 317.

[122] M. Bodner and A. Cadavid, "Dimensional reduction of type IIB supergravity and exceptional quaternionic manifolds," *Class.Quant.Grav.* **7** (1990) 829.

[123] M. Bodner, A. Cadavid, and S. Ferrara, "(2,2) vacuum configurations for type IIA superstrings: N=2 supergravity Lagrangians and algebraic geometry," *Class.Quant.Grav.* **8** (1991) 789–808.

[124] P. A. Dirac, "Quantized Singularities in the Electromagnetic Field," *Proc.Roy.Soc.Lond.* **A133** (1931) 60–72.

[125] R. I. Nepomechie, "Magnetic Monopoles from Antisymmetric Tensor Gauge Fields," *Phys.Rev.* **D31** (1985) 1921.

[126] C. Teitelboim, "Gauge Invariance for Extended Objects," *Phys.Lett.* **B167** (1986) 63.

[127] C. Teitelboim, "Monopoles of Higher Rank," *Phys.Lett.* **B167** (1986) 69.

[128] R. Rohm and E. Witten, "The Antisymmetric Tensor Field in Superstring Theory," *Ann. Phys.* **170** (1986) 454.

[129] C. Hull, "Superstring compactifications with torsion and space-time supersymmetry,".

[130] A. Strominger, "Superstrings with Torsion," *Nucl.Phys.* **B274** (1986) 253.

[131] B. de Wit, D. Smit, and N. Hari Dass, "Residual Supersymmetry of Compactified D=10 Supergravity," *Nucl.Phys.* **B283** (1987) 165.

[132] K. Becker and M. Becker, "M theory on eight manifolds," *Nucl.Phys.* **B477** (1996) 155–167, arXiv:hep-th/9605053 [hep-th].

[133] S. B. Giddings, S. Kachru, and J. Polchinski, "Hierarchies from fluxes in string compactifications," *Phys.Rev.* **D66** (2002) 106006, arXiv:hep-th/0105097 [hep-th].

[134] G. Lopes Cardoso et al., "Non-Kaehler string backgrounds and their five torsion classes," *Nucl. Phys.* **B652** (2003) 5–34, arXiv:hep-th/0211118.

[135] K. Becker, M. Becker, K. Dasgupta, and P. S. Green, "Compactifications of heterotic theory on nonKahler complex manifolds. 1.," *JHEP* **0304** (2003) 007, arXiv:hep-th/0301161 [hep-th].

[136] K. Becker, M. Becker, K. Dasgupta, and S. Prokushkin, "Properties of heterotic vacua from superpotentials," *Nucl.Phys.* **B666** (2003) 144–174, arXiv:hep-th/0304001 [hep-th].

[137] G. Lopes Cardoso, G. Curio, G. Dall'Agata, and D. Lüst, "BPS action and superpotential for heterotic string compactifications with fluxes," *JHEP* **0310** (2003) 004, arXiv:hep-th/0306088 [hep-th].

[138] G. Lopes Cardoso, G. Curio, G. Dall'Agata, and D. Lüst, "Heterotic string theory on nonKahler manifolds with H flux and gaugino condensate," *Fortsch.Phys.* **52** (2004) 483–488, arXiv:hep-th/0310021 [hep-th].

[139] K. Becker, M. Becker, P. S. Green, K. Dasgupta, and E. Sharpe, "Compactifications of heterotic strings on nonKahler complex manifolds. 2.," *Nucl.Phys.* **B678** (2004) 19–100, arXiv:hep-th/0310058 [hep-th].

[140] M. Becker and K. Dasgupta, "Kähler versus non Kähler compactifications," arXiv:hep-th/0312221 [hep-th].

[141] S. Gukov, C. Vafa, and E. Witten, "CFT's from Calabi-Yau four folds," *Nucl.Phys.* **B584** (2000) 69–108, arXiv:hep-th/9906070 [hep-th].

[142] T. R. Taylor and C. Vafa, "R R flux on Calabi-Yau and partial supersymmetry breaking," *Phys.Lett.* **B474** (2000) 130–137, arXiv:hep-th/9912152 [hep-th].

[143] K. Dasgupta, G. Rajesh, and S. Sethi, "M theory, orientifolds and G - flux," *JHEP* **9908** (1999) 023, arXiv:hep-th/9908088 [hep-th].

[144] S. Kachru, M. B. Schulz, and S. Trivedi, "Moduli stabilization from fluxes in a simple IIB orientifold," *JHEP* **0310** (2003) 007, arXiv:hep-th/0201028 [hep-th].

[145] S. Kachru, R. Kallosh, A. D. Linde, and S. P. Trivedi, "De Sitter vacua in string theory," *Phys.Rev.* **D68** (2003) 046005, arXiv:hep-th/0301240 [hep-th].

[146] J.-P. Derendinger, C. Kounnas, P. Petropoulos, and F. Zwirner, "Superpotentials in IIA compactifications with general fluxes," *Nucl.Phys.* **B715** (2005) 211–233, arXiv:hep-th/0411276 [hep-th].

[147] G. Villadoro and F. Zwirner, "N=1 effective potential from dual type-IIA D6/O6 orientifolds with general fluxes," *JHEP* **0506** (2005) 047, arXiv:hep-th/0503169 [hep-th].

[148] O. DeWolfe, A. Giryavets, S. Kachru, and W. Taylor, "Type IIA moduli stabilization," *JHEP* **0507** (2005) 066, arXiv:hep-th/0505160 [hep-th].

[149] P. G. Camara, A. Font, and L. Ibanez, "Fluxes, moduli fixing and MSSM-like vacua in a simple IIA orientifold," *JHEP* **0509** (2005) 013, arXiv:hep-th/0506066 [hep-th].

[150] M. Grana, "Flux compactifications in string theory: A Comprehensive review," *Phys.Rept.* **423** (2006) 91–158, arXiv:hep-th/0509003 [hep-th].

[151] M. R. Douglas and S. Kachru, "Flux compactification," *Rev.Mod.Phys.* **79** (2007) 733–796, arXiv:hep-th/0610102 [hep-th].

[152] F. Denef, M. R. Douglas, and S. Kachru, "Physics of String Flux Compactifications," *Ann.Rev.Nucl.Part.Sci.* **57** (2007) 119–144, arXiv:hep-th/0701050 [hep-th].

[153] S. Kobayashi and K. Nomizu, *Foundations of Differential Geometry*. Wiley Classics Library. John Wiley & Sons, 2009.

[154] C. Nash and S. Sen, *Topology and Geometry for Physicists*. Dover books on mathematics. Dover Publications, 2011.

[155] D. Joyce, *Compact manifolds with special holonomy*. Oxford mathematical monographs. Oxford University Press, 2000.

[156] M. Nakahara, *Geometry, topology, and physics*. Graduate student series in physics. Institute of Physics Publishing, 2003.

[157] D. Lüst, "Compactification Of Ten-dimensional Superstring Theories Over Ricci Flat Coset Spaces," *Nucl. Phys.* **B276** (1986) 220.

[158] J. P. Gauntlett, N. Kim, D. Martelli, and D. Waldram, "Five-branes wrapped on SLAG three cycles and related geometry," *JHEP* **0111** (2001) 018, arXiv:hep-th/0110034 [hep-th].

[159] J. P. Gauntlett, D. Martelli, and D. Waldram, "Superstrings with intrinsic torsion," *Phys.Rev.* **D69** (2004) 086002, arXiv:hep-th/0302158 [hep-th].

[160] M. Grana, R. Minasian, M. Petrini, and A. Tomasiello, "Supersymmetric backgrounds from generalized Calabi-Yau manifolds," *JHEP* **0408** (2004) 046, arXiv:hep-th/0406137 [hep-th].

[161] M. Grana, R. Minasian, M. Petrini, and A. Tomasiello, "Generalized structures of N=1 vacua," *JHEP* **0511** (2005) 020, arXiv:hep-th/0505212 [hep-th].

[162] S. Weinberg, "Implications of Dynamical Symmetry Breaking," *Phys.Rev.* **D13** (1976) 974–996.

[163] S. Weinberg, "Implications of Dynamical Symmetry Breaking: An Addendum," *Phys.Rev.* **D19** (1979) 1277–1280. (For original paper see Phys.Rev.D13:974-996,1976).

[164] E. Gildener, "Gauge Symmetry Hierarchies," *Phys.Rev.* **D14** (1976) 1667.

[165] e. 't Hooft, Gerard, e. Itzykson, C., e. Jaffe, A., e. Lehmann, H., e. Mitter, P.K., et al., "Recent Developments in Gauge Theories. Proceedings, Nato Advanced Study Institute, Cargese, France, August 26 - September 8, 1979," *NATO Adv.Study Inst.Ser.B Phys.* **59** (1980) 1–438.

[166] M. Prasad and C. M. Sommerfield, "An Exact Classical Solution for the 't Hooft Monopole and the Julia-Zee Dyon," *Phys.Rev.Lett.* **35** (1975) 760–762.

[167] E. Bogomolny, "Stability of Classical Solutions," *Sov.J.Nucl.Phys.* **24** (1976) 449.

[168] E. Witten and D. I. Olive, "Supersymmetry Algebras That Include Topological Charges," *Phys.Lett.* **B78** (1978) 97.

[169] G. Gibbons and C. Hull, "A Bogomolny Bound for General Relativity and Solitons in N=2 Supergravity," *Phys.Lett.* **B109** (1982) 190.

[170] A. Sen, "Strong - weak coupling duality in four-dimensional string theory," *Int.J.Mod.Phys.* **A9** (1994) 3707–3750, arXiv:hep-th/9402002 [hep-th].

[171] K. Landsteiner, E. Lopez, and D. A. Lowe, "Evidence for S duality in N=4 supersymmetric gauge theory," *Phys.Lett.* **B387** (1996) 300–303, arXiv:hep-th/9606146 [hep-th].

[172] M. Bershadsky, C. Vafa, and V. Sadov, "D-branes and topological field theories," *Nucl.Phys.* **B463** (1996) 420–434, arXiv:hep-th/9511222 [hep-th].

[173] J. P. Derendinger, L. E. Ibanez, and H. P. Nilles, "On the Low-Energy d = 4, N=1 Supergravity Theory Extracted from the d = 10, N=1 Superstring," *Phys. Lett.* **B155** (1985) 65.

[174] M. Dine, R. Rohm, N. Seiberg, and E. Witten, "Gluino Condensation in Superstring Models," *Phys.Lett.* **B156** (1985) 55.

[175] S. Ferrara, N. Magnoli, T. R. Taylor, and G. Veneziano, "Duality and supersymmetry breaking in string theory," *Phys. Lett.* **B245** (1990) 409–416.

[176] H. P. Nilles and M. Olechowski, "Gaugino condensation and duality invariance," *Phys. Lett.* **B248** (1990) 268–272.

[177] L. E. Ibanez and D. Lüst, "Duality anomaly cancellation, minimal string unification and the effective low-energy Lagrangian of 4-D strings," *Nucl. Phys.* **B382** (1992) 305–364, arXiv:hep-th/9202046.

[178] V. S. Kaplunovsky and J. Louis, "Model independent analysis of soft terms in effective supergravity and in string theory," *Phys. Lett.* **B306** (1993) 269–275, arXiv:hep-th/9303040.

[179] C. Bachas, "A Way to break supersymmetry," arXiv:hep-th/9503030 [hep-th].

[180] J. Polchinski and A. Strominger, "New vacua for type II string theory," *Phys.Lett.* **B388** (1996) 736–742, arXiv:hep-th/9510227 [hep-th].

[181] S. Gukov, "Solitons, superpotentials and calibrations," *Nucl.Phys.* **B574** (2000) 169–188, arXiv:hep-th/9911011 [hep-th].

[182] P. Mayr, "On supersymmetry breaking in string theory and its realization in brane worlds," *Nucl.Phys.* **B593** (2001) 99–126, arXiv:hep-th/0003198 [hep-th].

[183] G. Curio, A. Klemm, D. Lüst, and S. Theisen, "On the vacuum structure of type II string compactifications on Calabi-Yau spaces with H fluxes," *Nucl.Phys.* **B609** (2001) 3–45, arXiv:hep-th/0012213 [hep-th].

[184] M. Haack and J. Louis, "M theory compactified on Calabi-Yau fourfolds with background flux," *Phys.Lett.* **B507** (2001) 296–304, arXiv:hep-th/0103068 [hep-th].

[185] K. Becker and M. Becker, "Supersymmetry breaking, M theory and fluxes," *JHEP* **0107** (2001) 038, arXiv:hep-th/0107044 [hep-th].

[186] J. Louis and A. Micu, "Heterotic string theory with background fluxes," *Nucl.Phys.* **B626** (2002) 26–52, arXiv:hep-th/0110187 [hep-th].

[187] J. Louis and A. Micu, "Type 2 theories compactified on Calabi-Yau threefolds in the presence of background fluxes," *Nucl.Phys.* **B635** (2002) 395–431, arXiv:hep-th/0202168 [hep-th].

[188] K. Becker and K. Dasgupta, "Heterotic strings with torsion," *JHEP* **0211** (2002) 006, arXiv:hep-th/0209077 [hep-th].

[189] K. Becker, M. Becker, J.-X. Fu, L.-S. Tseng, and S.-T. Yau, "Anomaly cancellation and smooth non-Kahler solutions in heterotic string theory," *Nucl.Phys.* **B751** (2006) 108–128, arXiv:hep-th/0604137 [hep-th].

[190] K. Becker, C. Bertinato, Y.-C. Chung, and G. Guo, "Supersymmetry breaking, heterotic strings and fluxes," *Nucl.Phys.* **B823** (2009) 428–447, arXiv:0904.2932 [hep-th].

[191] R. Harvey and J. Lawson, H.B., "Calibrated geometries," *Acta Math.* **148** (1982) 47.

[192] J. Gutowski and G. Papadopoulos, "AdS calibrations," *Phys.Lett.* **B462** (1999) 81–88, arXiv:hep-th/9902034 [hep-th].

[193] J. Gutowski, G. Papadopoulos, and P. Townsend, "Supersymmetry and generalized calibrations," *Phys.Rev.* **D60** (1999) 106006, arXiv:hep-th/9905156 [hep-th].

[194] J. Evslin and L. Martucci, "D-brane networks in flux vacua, generalized cycles and calibrations," *JHEP* **0707** (2007) 040, arXiv:hep-th/0703129 [HEP-TH].

[195] P. Koerber and L. Martucci, "Deformations of calibrated D-branes in flux generalized complex manifolds," *JHEP* **0612** (2006) 062, arXiv:hep-th/0610044 [hep-th].

[196] L. Martucci and P. Smyth, "Supersymmetric D-branes and calibrations on general N=1 backgrounds," *JHEP* **0511** (2005) 048, arXiv:hep-th/0507099 [hep-th].

[197] D. Lüst, F. Marchesano, L. Martucci, and D. Tsimpis, "Generalized non-supersymmetric flux vacua," *JHEP* **0811** (2008) 021, arXiv:0807.4540 [hep-th].

[198] A. R. Frey and M. Lippert, "AdS strings with torsion: Non-complex heterotic compactifications," *Phys.Rev.* **D72** (2005) 126001, arXiv:hep-th/0507202 [hep-th].

[199] A. Bilal, J.-P. Derendinger, and R. Sauser, "M theory on S**1 / Z(2): Facts and fakes," *Nucl.Phys.* **B576** (2000) 347–374, arXiv:hep-th/9912150 [hep-th].

[200] E. Bergshoeff and M. de Roo, "The quartic effective action of the heterotic string and supersymmetry," *Nucl.Phys.* **B328** (1989) 439.

[201] C. Hull, "Anomalies, ambiguities and superstrings," *Phys.Lett.* **B167** (1986) 51.

[202] C. Hull, "Compactifications of the heterotic superstring," *Phys.Lett.* **B178** (1986) 357.

[203] D. Andriot, "Heterotic string from a higher dimensional perspective," *Nucl.Phys.* **B855** (2012) 222–267, arXiv:1102.1434 [hep-th].

[204] P. Koerber, "Lectures on Generalized Complex Geometry for Physicists," *Fortsch.Phys.* **59** (2011) 169–242, arXiv:1006.1536 [hep-th].

[205] J. P. Gauntlett and S. Pakis, "The Geometry of D = 11 killing spinors," *JHEP* **0304** (2003) 039, arXiv:hep-th/0212008 [hep-th].

[206] S. Chiossi and S. Salamon, "The Intrinsic torsion of SU(3) and G(2) structures," *J.Diff.Geom.* (2002), arXiv:math/0202282 [math-dg]. To Antonio Naveira on the occasion of his 60th birthday.

[207] G. Dall'Agata and N. Prezas, "N = 1 geometries for M theory and type IIA strings with fluxes," *Phys.Rev.* **D69** (2004) 066004, arXiv:hep-th/0311146 [hep-th].

[208] K. Behrndt, M. Cvetic, and T. Liu, "Classification of supersymmetric flux vacua in M theory," *Nucl.Phys.* **B749** (2006) 25–68, arXiv:hep-th/0512032 [hep-th].

[209] D. Huybrechts, *Complex geometry: an introduction*. Universitext (1979). Springer, 2005.

[210] R. L. Bryant, "Some remarks on G(2)-structures," arXiv:math/0305124 [math-dg].

[211] P. Kaste, R. Minasian, and A. Tomasiello, "Supersymmetric M theory compactifications with fluxes on seven-manifolds and G structures," *JHEP* **0307** (2003) 004, arXiv:hep-th/0303127 [hep-th].

[212] K. Behrndt and C. Jeschek, "Fluxes in M theory on seven manifolds: G structures and superpotential," *Nucl.Phys.* **B694** (2004) 99–114, arXiv:hep-th/0311119 [hep-th].

[213] K. Behrndt and C. Jeschek, "Fluxes in M theory on seven manifolds and G structures," *JHEP* **0304** (2003) 002, arXiv:hep-th/0302047 [hep-th].

[214] K. Behrndt and C. Jeschek, "Fluxes in M-theory on 7-manifolds: G(2-), SU(3) and SU(2)-structures," arXiv:hep-th/0406138 [hep-th].

[215] T. House and A. Micu, "M-Theory compactifications on manifolds with G(2) structure," *Class.Quant.Grav.* **22** (2005) 1709–1738, arXiv:hep-th/0412006 [hep-th].

[216] A. Micu, E. Palti, and P. Saffin, "M-theory on seven-dimensional manifolds with SU(3) structure," *JHEP* **0605** (2006) 048, arXiv:hep-th/0602163 [hep-th].

[217] L. Anguelova and K. Zoubos, "Flux superpotential in heterotic M-theory," *Phys.Rev.* **D74** (2006) 026005, arXiv:hep-th/0602039 [hep-th].

[218] L. Bedulli and L. Vezzoni, "The Ricci tensor of SU(3)-manifolds," *Journal of Geometry and Physics* **57** (Mar., 2007) 1125–1146, arXiv:math/0606786.

[219] D. Cassani, "Reducing democratic type II supergravity on SU(3) x SU(3) structures," *JHEP* **0806** (2008) 027, arXiv:0804.0595 [hep-th].

[220] M. Michelsohn, "On the existence of special metrics in complex geometry.," *Acta Math.* **149** (1982) 261–295.

[221] A. Sen, "(2, 0) Supersymmetry and Space-Time Supersymmetry in the Heterotic String Theory," *Nucl.Phys.* **B278** (1986) 289.

[222] R. L. Bryant, "Remarks on the geometry of almost complex 6-manifolds," *ArXiv Mathematics e-prints* (Aug., 2005), arXiv:math/0508428.

[223] P. Koerber and L. Martucci, "D-branes on AdS flux compactifications," *JHEP* **0801** (2008) 047, arXiv:0710.5530 [hep-th].

[224] E. Witten, "New Issues in Manifolds of SU(3) Holonomy," *Nucl.Phys.* **B268** (1986) 79.

[225] S. Donaldson, "Anti self-dual Yang-Mills connections over complex algebraic surfaces and stable vector bundles," *Proc.Lond.Math.Soc.* **50** (1985) 1–26.

[226] Y. S. Uhlenbeck, K., "On the existence of hermitian-yang-mills connections in stable vector bundles," *Communications on Pure and Applied Mathematics* **39** (1986) 257293.

[227] J. Li and S.-T. Yau, "Hermitian Yang-Mills connection on non-Kähler manifolds," *Conf.Proc.* **C8607214** (1986) 560–573.

[228] P. Griffiths and J. Harris, *Principles of Algebraic Geometry*. Wiley Classics Library Wiley Classics Library. John Wiley & Sons, 2011.

[229] F. Harvey and H. Lawson, "An introduction to potential theory in calibrated geometry," *Am.J.Math.* **131** (2009) 893–944.

[230] R. Abraham and J. Marsden, *Foundations of Mechanics*. AMS Chelsea publishing. AMS Chelsea Pub./American Mathematical Society, 1978.

[231] E. Cremmer, S. Ferrara, C. Kounnas, and D. V. Nanopoulos, "Naturally Vanishing Cosmological Constant in N=1 Supergravity," *Phys.Lett.* **B133** (1983) 61.

[232] A. Lahanas and D. V. Nanopoulos, "The Road to No Scale Supergravity," *Phys.Rept.* **145** (1987) 1.

[233] S. Gurrieri, A. Lukas, and A. Micu, "Heterotic String Compactifications on Half-flat Manifolds. II.," *JHEP* **0712** (2007) 081, arXiv:0709.1932 [hep-th].

[234] I. Benmachiche, J. Louis, and D. Martinez-Pedrera, "The Effective action of the heterotic string compactified on manifolds with SU(3) structure," *Class.Quant.Grav.* **25** (2008) 135006, arXiv:0802.0410 [hep-th].

[235] S. Gurrieri, A. Lukas, and A. Micu, "Heterotic on half-flat," *Phys.Rev.* **D70** (2004) 126009, arXiv:hep-th/0408121 [hep-th].

[236] B. de Carlos, S. Gurrieri, A. Lukas, and A. Micu, "Moduli stabilisation in heterotic string compactifications," *JHEP* **0603** (2006) 005, arXiv:hep-th/0507173 [hep-th].

[237] A. Micu, "A Note on Moduli Stabilisation in Heterotic Models in the Presence of Matter Fields," *Phys.Lett.* **B674** (2009) 139–142, arXiv:0812.2172 [hep-th].

[238] A. Micu, "Moduli Stabilisation in Heterotic Models with Standard Embedding," *JHEP* **1001** (2010) 011, arXiv:0911.2311 [hep-th].

[239] P. Koerber and L. Martucci, "From ten to four and back again: How to generalize the geometry," *JHEP* **0708** (2007) 059, arXiv:0707.1038 [hep-th].

[240] P. G. Camara and M. Grana, "No-scale supersymmetry breaking vacua and soft terms with torsion," *JHEP* **0802** (2008) 017, arXiv:0710.4577 [hep-th].

[241] J.-X. Fu and S.-T. Yau, "The theory of superstring with flux on non-Kaehler manifolds and the complex Monge-Ampere equation," *J. Diff. Geom.* **78** (2009) 369–428, arXiv:hep-th/0604063.

[242] K. Becker and S. Sethi, "Torsional Heterotic Geometries," *Nucl. Phys.* **B820** (2009) 1–31, arXiv:0903.3769 [hep-th].

[243] P. S. Aspinwall, "K3 surfaces and string duality," arXiv:hep-th/9611137 [hep-th].

[244] S. Gukov, S. Kachru, X. Liu, and L. McAllister, "Heterotic moduli stabilization with fractional Chern-Simons invariants," *Phys.Rev.* **D69** (2004) 086008, arXiv:hep-th/0310159 [hep-th].

[245] G. Curio, A. Krause, and D. Lüst, "Moduli stabilization in the heterotic/IIB discretuum," *Fortsch.Phys.* **54** (2006) 225–245, arXiv:hep-th/0502168 [hep-th].

[246] W. Fulton, *Introduction to toric varieties*. Annals of mathematics studies. Princeton University Press, 1993.

[247] M. Larfors, D. Lüst, and D. Tsimpis, "Flux compactification on smooth, compact three-dimensional toric varieties," *JHEP* **1007** (2010) 073, arXiv:1005.2194 [hep-th].

[248] N. Hitchin, "Generalized Calabi-Yau manifolds," *Quart.J.Math.Oxford Ser.* **54** (2003) 281–308, arXiv:math/0209099 [math-dg]. 37 pages, LateX Subj-class: Differential Geometry: Algebraic Geometry MSC-class: 53C15, 53C80, 53D30.

[249] M. Gualtieri, "Generalized complex geometry," arXiv:math/0401221 [math-dg]. Ph.D. Thesis (Advisor: Nigel Hitchin).

[250] E. Bergshoeff and M. de Roo, "Duality transformations of string effective actions," *Phys.Lett.* **B249** (1990) 27–34.

[251] U. Gran, "GAMMA: A Mathematica package for performing Gamma-matrix algebra and Fierz transformations in arbitrary dimensions," arXiv:hep-th/0105086.

i want morebooks!

Buy your books fast and straightforward online - at one of world's fastest growing online book stores! Environmentally sound due to Print-on-Demand technologies.

Buy your books online at
www.get-morebooks.com

Kaufen Sie Ihre Bücher schnell und unkompliziert online – auf einer der am schnellsten wachsenden Buchhandelsplattformen weltweit! Dank Print-On-Demand umwelt- und ressourcenschonend produziert.

Bücher schneller online kaufen
www.morebooks.de

VDM Verlagsservicegesellschaft mbH
Heinrich-Böcking-Str. 6-8   Telefon: +49 681 3720 174   info@vdm-vsg.de
D - 66121 Saarbrücken   Telefax: +49 681 3720 1749   www.vdm-vsg.de

Printed by Books on Demand GmbH, Norderstedt / Germany